Everyday Statistical Reasoning

Possibilities and Pitfalls

Everyday Statistical Reasoning
Possibilities and Pitfalls

Timothy J. Lawson
College of Mount St. Joseph

Australia • Canada • Mexico • Singapore • Spain • United Kingdom • United States

Publisher: *Vicki Knight*
Editorial Assistants: *Julie Dillemuth,*
 Dan Moneypenny
Marketing Manager: *Joanne Terhaar*
Marketing Assistant: *Justine Ferguson*
Production Editor: *Kirk Bomont*
Production Service: *Scratchgravel Publishing Services*
Manuscript Editor: *Carol Lombardi*

Permissions Editor: *Sue Ewing*
Interior Design: *Lisa Devenish*
Cover Design: *Denise Davidson*
Cover Art: *Ryuichi Okano/Photonic*
Interior Illustration: *Anne Draus*
Print Buyer: *Kris Waller*
Typesetting: *Scratchgravel Publishing Services*
Printing and Binding: *Webcom Limited*

COPYRIGHT © 2002 Wadsworth Group. Wadsworth is an imprint of the Wadsworth Group, a division of Thomson Learning, Inc. Thomson Learning™ is a trademark used herein under license.

For more information about this or any other Wadsworth products, contact:
WADSWORTH
511 Forest Lodge Road
Pacific Grove, CA 93950 USA
www.wadsworth.com
1-800-423-0563 (Thomson Learning Academic Resource Center)

All rights reserved. No part of this work covered by the copyright hereon may be reproduced or used in any form or by any means—graphic, electronic, or mechanical, including photocopying, recording, taping, Web distribution, or information storage and retrieval systems—without the written permission of the publisher.

For permission to use material from this work, contact us by
Web: www.thomsonrights.com
Fax: 1-800-730-2215
Phone: 1-800-730-2214

Printed in Canada

10 9 8 7 6 5 4 3 2 1

Library of Congress Cataloging-in-Publication Data

Lawson, Timothy J., [date]–
   Everyday statistical reasoning : possibilities & pitfalls / Timothy J. Lawson.
       p. cm.
   Includes bibliographical references and index.
   ISBN 0-534-59094-2 (alk. paper)
     1. Social sciences—Statistical methods.   I. Title.

HA29 .L28 2002
001.4'22—dc21
                                                            2001026018

To Anna, Alexandra, and Ryan

About the Author

Timothy J. Lawson is professor of psychology at the College of Mount St. Joseph in Cincinnati, Ohio. He received his master's and Ph.D. in social psychology from Miami University. His teaching interests include statistics, research methods, social psychology, social influence, senior thesis, and introductory psychology. Dr. Lawson has published in a variety of professional journals in the areas of attribution, humor, lying, and the teaching of psychology. He is a member of the American Psychological Society, Society for Personality and Social Psychology, Society for the Teaching of Psychology, Midwestern Psychological Association, and Council of Teachers of Undergraduate Psychology (CTUP). His personal interests include golf and cooking various ethnic foods.

Contents

1 **Statistics Aren't Just for Research Anymore / 1**
Questionnaire / 3

2 **Probability: Some Basic Rules / 7**
Probability and Research / 7
Subjective Probability and Assessing Risk / 8
Using Heuristics to Assess Risk / 10
Conjunction Junction / 13
The Conjunction Fallacy / 16
Probability and Chance Events / 18
Amazing Coincidences / 19
80% Certain, but 0% Sure / 21
Summary / 24
• PRACTICE PROBLEMS / 24

3 **The Law of Large Numbers / 25**
What Is the Law of Large Numbers? / 26
How Good Are We at Applying the LLN? / 28
The Representativeness Heuristic and Failures to Apply the LLN / 28
Do People Have an Intuitive Version of the LLN? / 29
The Impact of Single Cases / 32
Single Cases Versus Group Statistics / 32
"Person-who" Statistics / 33
Perceptions of Randomness / 35
Randomness and Belief in the "Hot Hand" / 37
Summary / 39
• PRACTICE PROBLEMS / 39

4 Estimation of Population Parameters: The Problem of Sample Bias / 43

The Process of Estimation / 44
Biased Samples from the Mass Media / 45
Attribution: Determining Cause / 47
Estimation Based on Single Cases / 49
Generalizing from the Self to Others / 51
The False Consensus Effect / 52
The Illusion of Transparency / 53
The Spotlight Effect / 55
Summary / 57
• PRACTICE PROBLEMS / 57

5 Correlation / 59

What Is a Correlation? / 61
Assessing Correlations in Everyday Life / 61
Difficulties in Assessing Correlations / 63
Assessing Correlations from Incomplete Data / 63
The Positive-Test Strategy / 64
Biased Expectations and Illusory Correlations / 66
Distinctive Events, Illusory Correlations, and Stereotypes / 69
Illusions of Control / 70
Accuracy of Correlation Detection / 71
Summary / 72
• PRACTICE PROBLEMS / 73

6 Regression and Prediction / 75

Correlation, Prediction, and Regression / 76
Regression Toward the Mean / 79
How Good Are We at Making Predictions? / 81
Regression Toward the Mean in Everyday Life / 84
Examples from the World of Sports / 84
Ozone and Gas Prices / 86
The Magazine Curse / 86
Alternative Health Remedies / 87
Other Ways to Conceptualize Regression Toward the Mean / 87
The Law of Large Numbers and Regression Toward the Mean / 88
Summary / 89
• PRACTICE PROBLEMS / 90

7 Factorial Analysis of Variance and Attribution / 93

What Is a Factorial Analysis of Variance? / 93
How Is a Factorial ANOVA Related to Everyday Reasoning? / 95
Single Versus Multiple Causes / 95
Attribution Theory / 97
Biased Attributions / 98
Summary / 100
• PRACTICE PROBLEMS / 101

8 Analysis of Covariance / 103

What Is Analysis of Covariance? / 104
Analysis of Covariance in Everyday Life / 105
ANCOVA and Stereotyping / 106
ANCOVA and Perceptions of Changes in Subpopulations / 109
Summary / 111
• PRACTICE PROBLEMS / 112

9 Conditional Probability and Bayes' Theorem / 113

Understanding Conditional Probability and Bayes' Theorem / 114
Common Difficulties in Bayesian Reasoning / 116
Confusion About Conditional Probabilities / 117
Base-Rate Neglect / 118
Improving Our Ability to Engage in Bayesian Reasoning / 120
Bayesian Reasoning in Legal Settings / 121
Conditional Probability and the Monty Hall Dilemma / 122
Summary / 124
• PRACTICE PROBLEMS / 124

Conclusion / 127

Glossary / 129

References / 133

Index / 141

Preface

The main purpose of this book is to explain how a number of statistical concepts can be applied to everyday decisions and judgments (the "possibilities" mentioned in the subtitle of this book). To accomplish this task, I rely heavily on scientific research—mainly psychological research—on human decision making, cognition, heuristics, biases, social perception, and stereotypes. This research highlights shortcomings in people's everyday statistical reasoning (the "pitfalls" in the subtitle), but it also demonstrates that people can improve their statistical reasoning with the proper training. Because of the emphasis on statistical reasoning and psychological research, this book is an ideal supplement for a wide range of courses, including those covering statistics, research methods, cognitive psychology, human decision making, and critical thinking.

Students who take college courses in statistics are often taught how to perform statistical procedures and how to use them to analyze and interpret data from research studies. Although the statistical *concepts* related to these procedures are also useful for making decisions and judgments in everyday life, many statistics textbooks do not emphasize this fact. That usefulness, however, is the main focus of this book.

In each chapter, I invite students to imagine themselves in everyday situations that require judgments, predictions, conclusions, or decisions that would benefit from proper statistical reasoning. I then explain how specific statistical concepts can be applied in those situations. Most of the statistical concepts (e.g., law of large numbers, probability, correlation, regression) are often covered in undergraduate statistics courses, but some (e.g., Bayes' theorem, analysis of covariance) are more likely to be found in graduate-level statistics courses. Because I explain these concepts in a conceptual manner with very little emphasis on formulas, a broad range of students—from undergraduates taking their first statistics course to graduate students who have had several statistics courses—should find this material understandable, interesting, and informative.

To make it easy for instructors to relate this material to topics that students are learning (or have already learned) in statistics courses, the chapters are named after common statistical principles or procedures. However, the chapters are not designed to teach students how to perform these statistical procedures. The main goal of each chapter is to show students how certain concepts associated with a particular statistical procedure (or principle) are useful in making everyday decisions and judgments.

To give students an opportunity to apply what they have learned, practice problems are included at the end of each chapter. These problems may be used for in-class exercises, homework assignments, or exam questions. Instructors may request the answers to these problems from Wadsworth–Thomson Learning.

Although it does not give extensive attention to the use of statistics in research, this book should broaden students' understanding of the statistical concepts that are central to the conduct of research in a variety of disciplines. I received training in the use of statistics in research during my undergraduate and graduate education and have used statistics in a variety of research studies. Nevertheless, I have found that learning how the concepts discussed in this book are relevant to everyday situations, decisions, and judgments has enriched my understanding of these statistical concepts.

I invite you to send me your personal examples of statistical reasoning in everyday life (that is, examples related to the statistical concepts or reasoning mistakes discussed in this book). I hope to incorporate these examples into future editions of this book. Please send your examples to Dr. Timothy J. Lawson, Psychology Program, College of Mount St. Joseph, 5701 Delhi Road, Cincinnati, OH 45233.

I thank Wallace Dixon and Hank Cetola for their encouragement and helpful comments on the first two chapters during the initial development of this book. Vicki Knight, my editor at Wadsworth, provided valuable ideas and guidance throughout the development process. Finally, I appreciate all of the praise and helpful suggestions provided by the following reviewers: Kevin Apple, James Madison University; Joan Ballard, SUNY at Geneseo; Patrick Conley, University of Illinois at Chicago; and David W. Martin, North Carolina State University.

Enjoy the book!

Timothy J. Lawson

Everyday Statistical Reasoning

Possibilities and Pitfalls

Statistics Aren't Just for Research Anymore

- Literacy and informed decision making in today's society require the ability to reason statistically.

  Derry, Levin, and Schauble (1995)

- If introductory statistics courses were to incorporate examples of how statistical principles such as the law of large numbers can be applied to judgments in everyday life, we have no doubt that such courses would have a more far-reaching effect on the extent to which people think statistically about the world.

  Fong, Krantz, and Nisbett (1986)

- Imagine that you recently saw a television news segment about your community blood centers running low on blood. Wanting to help, you decide to donate your blood. You drive to your local blood center, fill out a health history questionnaire on which you indicate that you are quite healthy, donate your blood, and go home. Prior to using your blood, the center performs a number of tests to determine if it contains particular viruses associated with infectious diseases [e.g., hepatitis, syphilis, human immunodeficiency virus (HIV)]. One day, while enjoying a candlelight dinner with your romantic partner, you receive a telephone call that changes your life. You learn that you tested positive for the AIDS virus (HIV). Feeling completely shocked and devastated, you quickly tell your partner you have to leave, and you drive to the blood center to speak with a physician. "There must be some mistake," you tell the physician. "I feel completely healthy, I have not had sex with anyone who has AIDS, and I have never used intravenous drugs." The physician responds, "The test is extremely accurate; if a person has HIV, the test detects it in 99.8% of the cases." "So," she continues, "I'm sorry, but the probability is practically 100% that you have HIV." You drive home in tears, wondering what the future holds. Will your romantic partner, whom you were about to marry, leave you? Will you lose your job? How long will you live?

  After some careful reading on the topic of AIDS, you discover that the rate of HIV infection in people like yourself—individuals who do not engage in risky behaviors, such as intravenous drug use—is about 1 in 10,000. The low incidence of the disease suggests that it is highly improbable that a person like you would have HIV. However, you also learn that the physician was correct about the "accuracy" of the test; among individuals who have HIV, the test detects it in 99.8% of the cases. Finally, you discover that the test incorrectly produces a positive result in only .01% of individuals who do not have HIV.

Was the physician correct in concluding that the "probability is practically 100% that you have HIV?" Actually, this physician made a serious error in statistical reasoning. Did you notice the error? The physician should have told you that the probability you have HIV is approximately 50%. Thus, it is certainly possible that you do not have HIV; in other words, it is possible that the test result was a false positive. Chapter 9 will discuss how to arrive at the figure of 50%.

Although the above situation is fictional, it is far from impossible. Researchers have discovered that people, including physicians, make this type of mistake in statistical reasoning when interpreting the results of laboratory tests (e.g., Casscells, Shoenberger, & Graboys, 1978; Gigerenzer & Hoffrage, 1995; Gigerenzer, Hoffrage, & Ebert, 1998; Hoffrage & Gigerenzer, 1998). Mistakes of this sort could have tragic results. For example, individuals from low-risk groups who test positive for HIV after donating blood might commit suicide because they are convinced it is an absolute certainty that they have the virus (cf. Gigerenzer et al., 1998).

This example illustrates, in a rather dramatic way, the title of this chapter: Statistics aren't just for research anymore. In other words, the topic of statistics should not be approached as though it is relevant only to those who will someday conduct scientific research or read such research. Statistical concepts are also very useful for helping us make decisions, conclusions, and predictions in everyday life—an important point often underemphasized in textbooks on statistics and research methodology.

After many years of teaching statistics to psychology, sociology, social work, and other students, I have found that many of them approach statistics as something that they can forget after the course because they believe they will never use them again. For many of them, the main goal is to learn how to plug numbers into formulas (or statistical software) and arrive at the correct answer. In fact, some students—and even some of their professors—wonder why they need to learn statistics at all.

The fact is, whether we realize it or not, we often make decisions and judgments in our everyday lives that require statistical reasoning, and applying the statistical concepts taught in statistics courses to everyday problems can greatly enhance our decision-making and critical-thinking skills. Poor statistical reasoning can create problems in our lives. These problems might be fairly trivial (e.g., choosing a poor restaurant based on the recommendation of a friend who ate there once), somewhat serious (e.g., betting $1,000 that the next spin of the roulette wheel will land on black because the last four spins landed on red), or very serious (e.g., being misdiagnosed as having HIV or cancer based on a positive test result). Fortunately, psychologists have found that training in statistics and in the application of statistical concepts to everyday problems improves people's ability to use statistical reasoning with everyday events (see Fong, Krantz, & Nisbett, 1986; Nisbett, Krantz, Jepson, & Kunda, 1983). However, even people who have received extensive formal training in statistics—including people with doctoral degrees in medicine and psychology—exhibit some of the mistakes in everyday statistical reasoning that we will examine in this book. As Nisbett et al. implied, perhaps a portion of these mistakes are due to the fact that "courses in statistics do not emphasize ways to use statistical principles in everyday life" (p. 359). This book is designed to improve this situation.

This book will examine a number of statistical concepts—including probability, the law of large numbers, correlation, and regression—and discuss how they are related to everyday decisions, conclusions, and predictions. We will focus mainly on the concepts with which people have difficulty at times and discuss a number of heuristics (i.e., mental shortcuts), biases, and other cognitive illusions that cause or are caused by poor statistical reasoning. Along the way, you will learn about some fascinating psychological research on a variety of topics, including gambling, jury decision making, stereotyping, superstitious beliefs, intimate relationships, and our perceptions of others.

To make it easier to relate this material to statistics courses you have taken (or are taking), each chapter is named after a common statistical principle or procedure. However, unlike a typical statistics textbook, the main goal of each chapter is *not* to teach you how to perform statistical procedures or interpret the results of such procedures. The main goal is to show you how certain *concepts* related to each statistical procedure are useful for everyday decisions and judgments. Thus, you may notice that some of the examples involve the intuitive application of statistical concepts in situations where one does not have the proper data to perform the statistical *procedure* related to these concepts. The concepts may nevertheless be applied in these situations.

Before we get started, please complete the following questionnaire, which contains a number of problems that will be considered throughout this book. If you like suspense, you can wait until you read the chapter that contains each problem before you check your answers; if not, use the chapter number provided to find a more detailed discussion of each problem.

Questionnaire

Directions: *Please write your answers in the space provided after each question or use a separate piece of paper if you do not want to write in this book.*

1. **A.** If you were to become a victim of a crime, how likely is it that it would be a *violent* crime (e.g., rape, murder, or assault)? To assess this risk, you might first estimate what percentage of all crimes reported in the United States are violent crimes. What is your estimate of this percentage? _____

B. Do you know what diseases or events are frequent causes of death in the United States? For example, do you think it is more likely that a person's death would be caused by chronic liver disease or homicide? _____ Is it more likely that a person's death would be caused by cancer of the digestive system or a motor vehicle accident? _____ (Chapter 2).

2. You are thinking of buying a new computer but cannot decide which one to buy. You know that you want one that has the fastest microprocessor, the largest hard drive, and the best software included in the price. But several brands have these features, and you must choose among them. Wanting your computer to last at least several years, you decide to choose a brand that will be very reliable and have few problems. You visit your college's computer center and get some advice from the person in charge of buying their computers. She tells you that the college purchased computers several years ago. They bought Brand A computers to fill one of their large computer labs and Brand B computers to fill a second large computer lab. She also informs you that both brands have performed fairly well, but the Brand A computers have had somewhat fewer problems. While discussing computers with your friends, you learn that a good friend of yours bought a Brand A computer a couple of years ago and had nothing but trouble with it. Only two months after he purchased his computer, the hard drive failed. After the hard drive was replaced, the monitor died; once that was replaced, the memory chips failed. After talking with your other friends, you discover that one of them purchased a Brand B computer a couple of years ago and she absolutely loves it. Her computer has performed flawlessly, and she highly recommends you buy one. Which brand (A or B) would you choose? _____ (Chapter 3).

3. Several years ago, a student of mine came to my office to discuss the poor grade she received on an exam in my class. Her anger was clearly visible as she explained her belief that my exams were too difficult. She had reached this conclusion after talking with a couple of other students in the class who also performed poorly on the exam. If you were this student, would you reach the same conclusion? Would you be angry? Explain your answer. _____
_____
_____
(Chapter 4).

4. A man suspects that blond women tend to be especially talkative compared with other women. Assume that, over the period of a year, he met and talked to 40 blond women who were very talkative. Thus, he concludes that there is indeed a relationship—or correlation—between the attributes "blond" and "talkative." Do you think he is correct? _____ (Chapter 5).

5. You were a good student in high school and expect to do quite well in college. In your first college course, Introduction to Psychology, you decide to put forth your best effort. You read the textbook carefully every week, go to every class, and take copious notes. However, when it comes time to study for the first exam, you come down with the flu and are able to study only half as long as you would have liked. To your surprise, you obtain the highest possible score (100) on the exam. You feel elated, especially after being told that the class average was 72. You continue to work hard in the course. When it comes time to study for the second exam you study long and hard, spending much more time than you did for the first exam. Nevertheless, your score on the second exam drops to 92 (again, the class average was 72). How would you explain this result? Should you conclude that you studied too much? _____

_____

_____

(Chapter 6).

6. Imagine that you volunteer to participate in a psychological study. When you arrive for the experiment, you are ushered to a small room. You put on headphones and listen to tape-recorded instructions that explain that you will be participating in a discussion with five other students about personal problems faced by college students. You will not meet these students in person, but they will be just down the hall in individual rooms. In order to avoid embarrassment and keep the discussion anonymous, the discussion will take place over an intercom system. Each student is supposed to present his or her problems to the group, one at a time. The first student to begin the discussion explains that he has had some difficulty getting adjusted to the city and his studies. With some hesitation, he also explains that he sometimes has seizures. Then, you hear a number of other students discuss their problems. When the other students are finished, the first student begins talking again. He makes a few comments and then begins to get louder

and more incoherent. You hear him stutter and explain, in a way that is barely understandable, that he needs help because he is having a seizure. He continues to beg for help, then starts choking while saying that he is going to die. You hear him choke again, and then the intercom system cuts off his voice. What would you do in this situation? Would you leave your room and try to help this desperate student? _____ If I told you that a person was actually in this situation and he did not help, what would you think of him?_____

_____

Would you view him differently if I told you that his response was fairly typical of other people who were in this situation? _____

(Chapter 7).

7. Imagine that we tested the verbal intelligence of members of two different racial groups (Group A and Group B). Both groups were given 25 anagrams to solve, some of which were five-letter anagrams and some seven-letter anagrams. Members of Group A solved all 5 (100%) of the five-letter anagrams and 5 (25%) of the 20 seven-letter anagrams they were given. Members of Group B solved 15 (75%) of the 20 five-letter anagrams and none (0%) of the 5 seven-letter anagrams they were given. Considering their performance on the anagrams, which group would you say has the highest verbal intelligence? _____ (Chapter 8).

8. Suppose that you learned in your introductory psychology class that most of the people who have committed suicide talked about doing it beforehand. Would you infer from this information that a person who talks about suicide is very likely to commit suicide? Explain your answer. _____

_____

_____

(Chapter 9).

Probability: Some Basic Rules

> Probability, like logic, is not just for mathematicians anymore. It permeates our lives.
>
> *Paulos (1988)*

Suppose that you are searching for a financial consultant to help you determine which stocks would be good investments for increasing your wealth before you retire. While doing some research on the consultants in your area, you come across a magazine article that describes an investment contest that took place recently. One thousand financial consultants from across the country were asked to predict the performance of Proctor and Gamble (P & G) stock for 10 weeks in a row. Each week, the consultants simply stated whether they thought the price of P & G stock would increase or decrease that week. The consultant who won the contest made correct predictions 10 weeks in a row. In your local newspaper, you see an advertisement featuring this consultant, and you realize that her office is near your home. In the advertisement, she touts her success in the contest and says "Invest your money wisely . . . choose the financial consultant who recently beat 999 of the top consultants in the country." Does her performance in the contest seem like convincing evidence of her ability to predict the performance of stocks? Would you choose her as your consultant? Remember your answer, and we will come back to this example later in this chapter.

First, we will examine the concept of probability and briefly discuss how this concept is useful to researchers. Then, we will discuss how the concept of probability is relevant to some of the decisions and conclusions we make in our everyday lives, including our (a) assessment of risks, (b) estimation of the likelihood of combinations of events, and (c) perceptions of chance events. Along the way, we will examine some of the mental shortcuts—or **heuristics**—we sometimes use to estimate probabilities.

Probability and Research

The concept of probability pervades statistics and scientific research. Researchers often calculate probabilities in order to determine the likelihood that the results of an experiment occurred by chance. For instance, if a researcher finds that a small group of students who tried a new study technique scored higher on a test than did a control group of students, she would want to know the probability that the groups' test scores differed by chance. If it were very likely that the difference occurred by chance, she would not conclude that the new study technique was effective.

Consider another example of the use of probability in research. Suppose a researcher wants to test a man who claims to have the ability to read minds. After blindfolding him, the researcher obtains a new deck of 52 cards and randomly selects 1 card from the deck. Then, the researcher asks him to guess which card was selected. What is the probability that he will guess correctly by chance? You probably quickly determined that the probability is 1/52. This answer can be determined from the following basic formula commonly found in statistics textbooks:

$$\text{probability of A} = \frac{\text{number of outcomes classified as A}}{\text{total number of possible outcomes}}$$

Let's say A represents correctly guessing the card you chose. There is only one outcome (or card) classified as A, and there are a total of 52 possible outcomes (or 52 cards). Therefore, the probability of guessing correctly by chance is 1/52, or 1.9%.

It is very unlikely that a person would guess the correct card by chance, but let's assume the researcher wants to make it extremely unlikely that the man could pass her test by chance. She decides to have him perform this test twice. Thus, after he makes his first guess, she will put the card back in the deck, shuffle the cards, and randomly select a second card. What is the probability that the man will make two correct guesses by chance? This probability can be determined by the **multiplication rule**, which requires that we simply multiply the probability of the first correct guess by the probability of the second correct guess. For each guess, the probability of being correct by chance is 1/52, so the probability of two correct guesses is 1/52 × 1/52 or .04%. Thus, it is extremely unlikely that a person could make two correct guesses by chance. I should note that this multiplication rule should be used only when the two outcomes are independent of one another. In the card example, regardless of whether the man guesses correctly or incorrectly on the first try, the probability of a correct guess on the second try is 1/52. These two outcomes are independent, because what happens on the first try does not affect the probability of being correct on the second try.

Subjective Probability and Assessing Risk

As the quote at the beginning of the chapter stated, probability is not just for mathematicians or researchers. Although we may not carry around calculators to make precise calculations of probabilities, we do develop subjective estimates of probabilities, or **subjective probabilities**. You might be surprised by the number of decisions and conclusions we make in our daily lives that are based on our estimates of the probabilities of events. For example, your decision to take, avoid, or drop a particular college course might be influenced by your belief concerning how likely it is that you will pass the course. Whether you speed while driving down the highway might depend on your estimate of the probability that you will get caught by the police. Your decision to exercise or stop smoking might be based on your belief that these behaviors are likely to improve your health. Your conclusion that a person who harms someone is evil might be influenced by how likely you think it is that you (or somebody else) would have done the same thing if put in that situation. These are just a few of the many ways that probability permeates our lives.

One of the things we frequently need to do is assess the probability of various risks, such as the probability that we will get lung cancer if we smoke cigarettes or the probability we will be in a fatal crash if we fly rather than drive somewhere. Moreover, our decisions concerning what types of products or insurance policies to buy may be influenced by our estimates of various risks. For instance, your willingness to purchase an expensive security alarm system for your home may depend on your subjective probability of being the victim of a burglar or some other criminal. Let's take a look at your estimates of specific types of risks. Recall Question 1A from the questionnaire in Chapter 1:

> **A.** If you were to become a victim of a crime, how likely is it that it would be a *violent* crime (e.g., rape, murder, or assault)? To assess this risk, you might first estimate what percentage of all crimes reported in the United States are violent crimes. What is your estimate of this percentage?

If you are like most of my students, you probably overestimated this percentage. If, on the other hand, you guessed 12%, you are correct. According to the Federal Bureau of Investigation's (1998) report titled *Crime in the United States*, a total of 12,475,634 crimes were reported in the United States in 1998, and 1,531,044 of those crimes were classified as violent. Thus, if we were to randomly choose a person who had been a victim of a reported crime in 1998, the probability is 12.3% (1,531,044/12,475,634) that the crime would be a violent crime. You may have noticed that I used the probability formula given on page 8 to arrive at this answer (i.e., the probability of a crime being violent = number of violent crimes/total number of crimes). However, it would be misleading to suggest that this percentage is the probability that *you* will be involved in a violent crime in any particular year. One reason is that the percentage is based on the assumption that you were involved in a reported crime in the first place, and this was a rare event that happened to only about 5% of the population in 1998 (roughly .6% of the population were victims of a violent crime). Another reason is that the percentage does not take into account your age, sex, race, or the region of the country in which you live, and these factors could influence the probability that you would be a victim of crime. A third reason is that the figure is based on *reported* crimes, and many crimes go unreported. Although accurately predicting how likely it is that you will be a victim of a violent crime is a difficult task, many of us overestimate this probability. Being aware of this tendency might help bring our estimates closer to reality.

Let's revisit your estimate of another risk. Recall Question 1B from the questionnaire:

B. Do you know what diseases or events are frequent causes of death in the United States? For example, do you think it is more likely that a person's death would be caused by chronic liver disease or homicide? Is it more likely that a person's death would be caused by cancer of the digestive system or a motor vehicle accident?

If you chose the second cause in each question (i.e., homicide and motor vehicle accident), you are wrong. The correct answers are chronic liver disease and cancer of the digestive system.[1] According to the National Center for Health Statistics (see Hoyert, Kochanek, & Murphy, 1999), the 1997 death rates (per 100,000 inhabitants) were 9.4 for chronic liver disease and 7.4 for homicide. In other words, the number of people who died from chronic liver disease was 27% higher than the number who died from a homicide. Moreover, the 1997 death rates were 47.5 for cancer of the digestive system and 16.2 for motor vehicle accidents.[2] Thus, cancer of the digestive system killed almost three times as many people as did motor vehicle accidents! I find this latter statistic very surprising. Apparently, I am not alone. Lichtenstein, Slovic, Fischhoff, Layman, and Combs (1978) found that people make interesting errors in estimating the frequency of various causes of death. They asked college students and members of the League of Women Voters to provide estimates for 41 causes of death. They found that although their participants tended to give higher estimates for more-frequent causes of death, they also made some systematic errors. Their participants tended to overestimate the frequency of causes of death that were dramatic or sensationalized (e.g., accidents, tornadoes, floods, and homicide) and underestimate the frequency of causes that were less spectacular (e.g., diabetes, stomach cancer, stroke, and asthma).

Using Heuristics to Assess Risk

Why do people make these types of errors when judging the frequency of various causes of death? For example, why does it *seem* much more probable that someone's death would be a result of a motor vehicle accident as opposed to cancer of the digestive system? One reason is that, when we judge the probability of events such as these,

---

[1] I did not select the *most* frequent causes of death for these examples. In 1997, heart disease and cancer were the top two causes of death in the United States.

[2] All of the death rates in this paragraph are for all deaths regardless of age, sex, or race. Thus, as with the crime statistics considered earlier, these figures may not accurately portray the probability that *you* would die from one of these causes. Readers who would like to get a better sense of death rates specific to their age, sex, and race can consult the National Center for Health Statistics reports for more detailed information.

we may rely on how easily we can imagine or recall them happening. This cognitive shortcut is called the **availability heuristic** (Lichtenstein et al., 1978; Tversky & Kahneman, 1973).

Before we examine how the availability heuristic could lead to inaccurate judgments, note that this heuristic can be a useful tool for estimating probabilities. In general, events that are more frequent in our lives should be easier to recall, or more cognitively available, than less-frequent events. Thus, if we use the availability heuristic, our estimates of frequency or probability should increase as actual frequency or probability increases. For example, if you live in southern California, your estimate of the probability of experiencing an earthquake is likely to be higher than that of a person who lives in Michigan because earthquakes are more frequent in California than in Michigan.

However, the availability heuristic can lead to inaccurate judgments because the ease with which something comes to mind can be influenced by factors other than its actual frequency. For instance, famous names may be easier to recall than less-famous names, even if we hear the less-famous names more frequently. Tversky and Kahneman (1973, Study 8) investigated this possibility by having people listen to recorded lists of names of public figures and entertainers, some of which were very famous (e.g., Elizabeth Taylor) and some that were less-famous (e.g., William Fulbright). Half of the lists consisted of 19 names of famous women and 20 names of less-famous men, and the other half consisted of 19 names of famous men and 20 names of less-famous women. The researchers asked one group of participants to recall as many of the names as they could and another group to judge whether the list contained more names of men or of women. They reasoned that famous names would be easier for people to recall and that the increased availability of these names would distort participants' frequency judgments. This held true: Although the lists actually contained fewer famous names than less-famous names, the majority of participants recalled more famous names; they recalled an average of 12.3 of the 19 famous names and 8.4 of the 20 less-famous names. Moreover, the participants incorrectly estimated that (a) men's names were more frequent in the lists containing famous men and less-famous women and (b) women's names were more frequent in the lists containing famous women and less-famous men.

The availability heuristic can also distort our judgments of the frequency of various causes of death. How easily various causes of death come to mind when we are estimating frequencies is influenced by how often we hear about such causes from the mass media. If newspapers and television programs are more likely to report some causes of death than others, our judgments of the frequency of these causes may not match reality. Combs and Slovic (1979) examined newspaper reports of causes of death and found that dramatic or violent causes of death (e.g., accidents, tornadoes, and homicide) were reported much more frequently than were less-dramatic causes (e.g., diabetes, stroke, and stomach cancer). Does this sound familiar? The dramatic causes reported often by newspapers are also ones that we tend to overestimate in our frequency judgments, and the less-dramatic causes that are often underreported by newspapers are the ones we tend to underestimate in our frequency judgments.

Clearly, when it comes to estimating risks associated with various events—such as causes of death or crimes—we need to be aware that our subjective estimates could be flawed, especially when those events are dramatic or sensationalized. Overestimating the likelihood of an airplane crash, for example, could make you decide to drive rather than fly to your destination. This decision, however, would be a dangerous one, given that the probability you would die in a car accident is *much* higher—perhaps 26 times higher—than the probability you would die in an airplane accident (cf. National Safety Council, 1991).

Of course, your subjective probabilities of various misfortunes can be distorted by things other than the media. Personally experiencing an improbable misfortune or hearing about another person's experience with such an event might also distort your estimate of the probability of such an event. For example, a friend of mine told me that several years ago his parents were on a flight from Toledo, Ohio, to Las Vegas, Nevada. During the flight one of the airplane's engines exploded, and they were forced to make an emergency landing in Detroit. As you can imagine, this scared a lot of the passengers. Many of them refused to get on another airplane to finish the trip to Las Vegas. Does their refusal seem rational to you? You might be thinking, "Of course it does; I wouldn't get on another plane at that point even if you paid me to do it." This is understandable in light of our discussion of the availability heuristic. The very emotional, memorable experience of a mechanical failure on an airplane probably increased the passengers' subjective probability of experiencing a problem on a second flight. However, the *actual* probability that the passengers would experience a mechanical failure on a second airplane was probably much lower than it seemed to them at the time. Perhaps the most rational passengers were those who took the second airplane to Las Vegas. My friend's parents were among those passengers, and I am happy to report that they survived the flight quite well.

Another implication of our use of the availability heuristic when estimating probabilities is that the *recency* of an event could affect our judgment of its probability. A recent event may be particularly memorable and therefore seem very probable. As time passes and our memory fades, the event may seem less probable, even though the actual probability of the event has not changed during that time period. For example, immediately after experiencing a natural hazard (e.g., flood, tornado, or hurricane) you might perceive the event as much more probable than you would at a later point in time. As Slovic, Fischhoff, and Lichtenstein (1982) stated:

> Availability bias helps explain people's misconceptions and faulty decisions with regard to certain natural hazards. . . . The purchase of earthquake insurance increases sharply after a quake and then declines steadily as memories fade. One particularly important implication of the availability heuristic is that discussion of a low-probability hazard may increase its memorability and imaginability and hence its perceived riskiness, regardless of what the evidence indicates. (p. 465)

Similarly, experiencing an improbable accident might increase your concern that it will happen to you again, and your concern may decline as time passes and your memory of the event fades. For instance, during the 1999 Oktoberfest festival in downtown Cincinnati, a man drank 88 ounces of beer in about 20 minutes. A while later, he decided to drive home. He ended up driving through the barricades set up around the festival streets and proceeded to drive through the crowd of people at the festival. He injured about 30 people, some of whom experienced concussions or broken legs. Before a judge sentenced him to more than 14 years in prison, several of the victims told the judge that the accident left them with some serious emotional, as well as physical, scars. Several victims said that they were afraid to drive or even cross the street; one said, "I can't cross without obsessively turning my head to check over and over again" (Horn, 2000, p. A1). The availability heuristic may be at work distorting these victims' probability estimates. The actual probability that this victim will be hit by a car while crossing the street has not increased since the time of the accident; but, subjectively, such an accident *seems* much more probable to the victim because of the recent accident. Perhaps, as the victim's memory of the accident fades, the subjective probability of such an accident will decrease, and crossing the street without obsessive checking will become easier.

Conjunction Junction

We have discussed the probabilities associated with various diseases and accidents. Now consider a different question: What is the probability that you will *avoid* being the victim of a serious disease or accident? Paulos (1988) invited readers to imagine the following hypothetical probabilities:

> Not being killed in a car accident may be 99 percent certain, while 98 percent of us may avoid perishing in a household accident. Our chances of escaping lung disease may be 95 percent; dementia, 90 percent; cancer, 80 percent; and heart disease, 75 percent. (p. 23)

Although these percentages are hypothetical, and the list of potential misfortunes is by no means exhaustive, take a moment to estimate how likely it is that you would avoid all of these misfortunes (i.e., state your chances in the form of a percentage, as Paulos did). Are your chances good?

Estimating your chances of avoiding a number of misfortunes requires calculating the probability of a **conjunction,** or a combination of two or more events. Assume that the aforementioned misfortunes are independent; in other words, the occurrence of one misfortune does not affect the probability of another misfortune (e.g., whether you have lung disease would not affect that probability that you would be in a fatal car accident). If not being killed in a car accident and not perishing in a household accident are independent events, we can use the multiplication rule to determine the probability of the conjunction of these events (i.e., probability of not being in a car accident *and* not being in a household accident) by simply multiplying the probabilities of the individual events. The general formula that reflects this rule is $p(A \& B) = p(A)p(B)$, where A and B stand for any two events (and p stands for probability). Thus, the probability that we would avoid both types of accidents is 97% (.99 × .98). We have a very good chance of avoiding two of these unfortunate events, but the probability of avoiding all of them is slightly less than 50% (.99 × .98 × .95 × .90 × .80 × .75). As Paulos (1988) said, "It's a little anxiety-provoking, how this innocuous multiplication principle can make our mortality more vivid" (p. 24).

Did you overestimate the probability that you would avoid all of those unfortunate events? Consider a different situation. You have the opportunity to win $100 by choosing the most probable event from the following: (a) drawing one red marble from a bag that contains 50% red and 50% white marbles or (b) drawing a red marble seven times in succession (putting the marble back after it is drawn each time) from a bag that contains 90% red marbles and 10% white marbles. Which would you choose? Bar-Hillel (1973) found that people given a similar choice preferred to bet on the conjunctive event (i.e., drawing a red marble seven times in succession) rather than the simple event (i.e., drawing one red marble). But, the probability of the conjunctive event is 48% (.9 × .9 × .9 × .9 × .9 × .9 × .9), which is lower than the probability of the simple event (50%).

Tversky and Kahneman (1974) hypothesized that one reason why we tend to overestimate the probability of conjunctive events is that we use the probability of the initial event (e.g., 90% chance of obtaining a red marble the first time) as a starting point, or anchor, for our estimate and adjust our estimate insufficiently from that point. This strategy has been called the **anchor-adjustment heuristic.** In this case, we adjust the initial estimate downward because we know the probability of the conjunction has to be less than 90%, but our final estimate remains too high because our adjustment was insufficient. The next paragraph presents some situations that require adjusting similar estimates upward. But, once again, the adjustment is insufficient.

In an experiment designed to investigate the use of the anchor-adjustment heuristic, Tversky and Kahneman gave a group of high school students 5 seconds to estimate the following product: 8 × 7 × 6 × 5 × 4 × 3 × 2 × 1. They asked a second group to estimate another product: 1 × 2 × 3 × 4 × 5 × 6 × 7 × 8. Of course, 5 seconds was not enough time for the participants to multiply all of the numbers together, so they had to estimate the answer. Given that the answer (40,320) was the same for both groups, one might expect the two groups to give equivalent estimates. However, if the students used the product of the first few numbers (e.g., 8 × 7 × 6) as an initial estimate (or anchor) and insufficiently adjusted that estimate to obtain their answers, the first group should have given a much higher estimate than the second group. The students' predictions were consistent with the use of the anchor-adjustment heuristic; the first group's median estimate (2,250) was more than four times higher than the second group's median estimate (512).

People's tendency to overestimate the probability of conjunctive events has implications for their ability to accurately predict the probability of sequences of events in real life. For example, when I am planning a trip from Cincinnati to southern Michigan to see my family, I sometimes arrange to meet my sister for dinner in Toledo on my way up to Michigan. Planning this event is difficult because it requires a fairly precise estimate of the time it will take me to travel from Cincinnati to Toledo (so I can tell her what time to meet me for dinner). To calculate my estimate, I often break the trip into segments and estimate the time involved in each one. I estimate what time I will leave my house and how long it will take me to go from Cincinnati to Dayton, Dayton to Lima, and Lima to the restaurant in Toledo. Assume that I predict I will leave my house at 4:30 P.M. I also estimate that it will take me one hour to travel from Cincinnati to Dayton, one hour to travel from Dayton to Lima, and one hour to travel from Lima to the restaurant. Therefore, I would tell my sister to meet me at the restaurant at 7:30 P.M.

In order for me to make it to the restaurant on time, the whole sequence of events has to occur exactly as I predicted. Thus, this sequence has a conjunctive character: I have to leave on time, *and* travel from Cincinnati to Dayton in one hour, *and* travel from Dayton to Lima in one hour, *and* drive from Lima to the restaurant in one hour. Assume the probability of each of these events is 80%. The probability of each of these events is never 100% because of all the factors that could delay me at each point. For example, I often take my family on the trip, so leaving on time depends on how long it takes them to get ready to leave. Moreover, traffic jams, accidents, and road construction can delay my progress in any of the segments of the trip. If the probability of each event in the sequence is 80% and the events are independent of one another, then the probability of all of them occurring as planned is only 41% (.8 × .8 × .8 × .8).[3] Now you know why I usually arrive at the restaurant later than I predicted: I often overestimate the probability of this conjunctive event. I suppose that, after my family reads this confession, I will have to stop overestimating this probability and generate a more reasonable estimate of arrival time.

People can make mistakes similar to the travel-time example in other situations as well, such as when business executives estimate the probability of success of a new product or when jurors estimate the accuracy of a prosecutor's claim that a sequence of events led to a murder (cf. Tversky & Kahneman, 1974, 1983). In each of these situations, a series of events has to occur to arrive at the outcome. Although each event in the series may be very likely, the probability of the entire series could be quite low. For instance, if General Mills were to develop a new cereal targeted at young children, the success of that product would depend on a number of factors: Children would have to like the taste of the cereal; parents would have to be willing to buy the cereal; grocery store managers would have to be willing to put the cereal on their

[3] I assume the events are independent because I tend not to drive much faster than the speed limit. Thus, if I encounter a traffic jam in one segment of the trip, I tend not to make up for lost time by traveling well beyond the speed limit on another segment of the trip.

shelves; and the General Mills marketing team would have to continue to create product packaging and advertising that appeals to consumers throughout the life of the product. The probability of each of these factors may be fairly high, but the probability of all of them occurring would be lower. Being aware of this fact might prevent a business executive from needlessly wasting time and money on a product that is unlikely to make it past all of these hurdles.

The Conjunction Fallacy

Another situation in which we sometimes have difficulty estimating the probability of conjunctive events is illustrated by the following example, used in a study by Tversky and Kahneman (1983):

> A health survey was conducted in a representative sample of adult males in British Columbia of all ages and occupations. Mr. F. was included in the sample. He was selected by chance from the list of participants. Which of the following statements is more probable? (check one)
>
> ☐ Mr. F. has had one or more heart attacks.
> ☐ Mr. F. has had one or more heart attacks and he is over 55 years old. (p. 305)

Did you estimate that the second statement was more probable than the first? Most of the college students in Tversky and Kahneman's study chose the second statement, but the first statement *must* be more probable than the second one. As can be seen in Figure 2.1, everyone who has had a heart attack *and* is over 55 years old is a member of the larger group of people who have had a heart attack. Thus, the conjunction

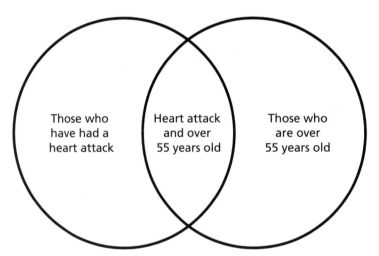

Figure 2.1 The populations of those who have had a heart attack and those who are over 55 years old

"heart attack *and* over 55 years old" cannot be more probable than "heart attack" alone. This is dictated by the **conjunction rule** of probability: $p(A \& B) \leq p(A)$, because A & B is a subset of A (where A stands for "heart attack" and B stands for "over 55 years old"). But, the conjunction *seems* more probable because a person who is over 55 years old fits well with our notion of someone who is likely to have a heart attack. Tversky and Kahneman used the term **"conjunction fallacy"** to describe such violations of the conjunction rule.

Consider another famous example used in a different study by Tversky and Kahneman (1983):

> Linda is 31 years old, single, outspoken and very bright. She majored in philosophy. As a student, she was deeply concerned with issues of discrimination and social justice, and also participated in anti-nuclear demonstrations. (p. 297)

Is it more likely that (a) Linda is a bank teller or (b) Linda is a bank teller and is active in the feminist movement? Given your knowledge of the conjunction rule, you know that the conjunction "bank teller *and* active in the feminist movement" cannot be more probable than "bank teller" alone. However, 85% of the college students who participated in this study indicated that the conjunction (statement b) was the most probable event.

Tversky and Kahneman (1983) wondered if participants could recognize the validity of the conjunction rule if they were presented with an argument for it, so they conducted another study similar to the "Linda" experiment. In addition to the description of Linda and statements a and b mentioned earlier, they gave participants the following two arguments and asked them to choose the one they found more convincing:

> **Argument 1.** Linda is more likely to be a bank teller than she is to be a feminist bank teller, because every feminist bank teller is a bank teller, but some women bank tellers are not feminists, and Linda could be one of them.
>
> **Argument 2.** Linda is more likely to be a feminist bank teller than she is likely to be a bank teller, because she resembles an active feminist more than she resembles a bank teller. (p. 299)

Once again, most (65%) of their participants flagrantly violated the conjunction rule by choosing Argument 2. Their choice of Argument 2 provides an interesting insight into why people commit the conjunction fallacy. People appear to be using a simple heuristic, called the **representativeness heuristic,** in which they rely on the degree to which the description of Linda is similar to (or representative of) either a bank teller or a bank teller who is active in the feminist movement. Because the description of Linda is more similar to the stereotype of a woman who would be involved in the feminist movement than it is to the stereotype of a bank teller, most people choose the former option. Tversky and Kahneman found additional evidence for this explanation in a variation on the Linda study in which participants were not given the full personality description of Linda; instead, they were told only that Linda was a 31-year-old woman. Participants in this study obeyed the conjunction rule and correctly indicated that it was more likely she was a bank teller than a bank teller who was active in the feminist movement.

Probability and Chance Events

Let's return to the financial example on page 7. Does the consultant's performance in the contest seem like convincing evidence of her ability to predict the performance of stocks? Would you choose her as your consultant?

Armed with the multiplication rule, you might decide to calculate how likely it is that the consultant would make 10 correct predictions by chance alone. If you assume that (a) the price of P & G stock was equally likely to increase or decrease during each of the 10 weeks and (b) the consultant was making random predictions each week, then she had a 50% chance of being correct each week. Thus, the probability that she would make 10 correct predictions by chance is slightly less than .1% ($.5^{10}$). This certainly appears to be convincing evidence of her amazing ability to predict stock performance. Are you convinced enough that you would choose her as your financial consultant?

Although your calculations of probability in the above example were correct, it is important to realize that, when given enough opportunity, improbable chance events are bound to occur at some point. This point is easier to understand if we look at the financial consultant example from a different perspective. Suppose that all 1,000 consultants who participated in the contest made their predictions by flipping a coin each week. If the coin landed on heads, an increase in stock price would be predicted; if it landed on tails, a decrease would be predicted. During the first week, imagine that half of the consultants predicted an increase in stock price, half predicted a decrease, and the stock price actually increased. Thus, 500 of the 1,000 consultants made a correct prediction. Because you are interested in accurate financial consultants, I will continue this example with only those consultants who made an accurate prediction. Of the 500 correct consultants, 250 predicted an increase the following week and 250 predicted a decrease, and the stock price increased again. During the third week, 125 of the 250 correct consultants predicted an increase, 125 predicted a decrease, and the stock price decreased. Continuing this trend would leave about 62 correct consultants (rounded to an even number) by the end of the fourth week, 31 after week five, about 16 after week six, 8 after week seven, 4 after week eight, 2 after week nine, and 1 "winner" after week ten. The winner's ten correct predictions would be due entirely to chance, but all of the potential investors who heard about it would be tempted to choose her as their personal financial consultant. As this example illustrates, we are sometimes tempted to attribute chance events to plausible causes when we should be attributing them to nothing other than chance.

Of course, the financial consultant contest is fictitious, but the general point of the example can be applied to situations that are quite real. As Stanovich (2001) and Fridson (1993) pointed out, it is not uncommon for investment advisors to win new clients by (a) pointing out the fact that they have beaten their peers several years in a row or (b) making a few correct predictions in a row. For at least some of these advisors, such outcomes are bound to occur by chance even if they make all of their predictions by simply flipping coins. Should we attribute some special skill to them? Clearly, the answer is no.

Amazing Coincidences

Another type of chance event for which people sometimes feel compelled to give causal or paranormal explanations is a coincidence. Surprising coincidences are more common than people think, but underestimating the likelihood of their occurrence can lead to inaccurate conclusions about the "cause" of such events. For example, on the day I was writing this section of the book, a coincidence occurred that my wife misattributed to scheming on my part. I was making plans for our holiday trip to Michigan, and I thought it would be nice for us to get together with one of my former professors, Hank, who lives in Adrian, Michigan. Hank agreed and suggested that we go to dinner at a restaurant in Ann Arbor that happened to be one of my favorite restaurants. This restaurant is one that I have taken my wife to on a number of previous trips to Michigan. In fact, she has been there so often that she has grown a bit tired of it. Nevertheless, she will often go to this restaurant if we are going with other people who want to eat there. Of course, when I informed her of Hank's suggestion, she became suspicious and said, "You want me to believe it was Hank's suggestion with no help from you?" It does seem awfully fishy, doesn't it? Hank's choice of restaurant seems improbable for a couple of reasons. First, despite the fact that Ann Arbor is roughly a 45-minute drive from Adrian, he chose a restaurant in Ann Arbor. Second, of all the restaurants in Ann Arbor he could have selected (there are many good ones), he chose one of my favorite restaurants. But, as I explained to my wife, I made no restaurant suggestions to him prior to his recommendation.

Do you see how overlooking the chance nature of coincidence can create problems (in the above example, it might lead to labeling me a liar or conniver)? A coincidence like this is not as uncommon as one might think. The facts are that Ann Arbor has far more excellent restaurants than does Adrian, and Hank has been to Ann Arbor many times; so Ann Arbor was a likely choice for a place to dine. Of course, there are many restaurants in Ann Arbor that Hank could have selected. But, he has had dinner with me a number of times and knows the type of food I like (in fact, my preferences are probably quite similar to his own), so this would narrow down the choices a bit. The choices would not even have to be narrowed down too much, because he could have selected a number of restaurants that would qualify as "one of my favorites." Thus, it is not that improbable that he would pick such a restaurant.

Other interesting coincidences are sometimes found among identical twins. Consider a pair of women: Both are Baptist, studying nursing in college, prefer vacations in historical locations, list tennis and volleyball as their favorite sports, and list English and math as their favorite subjects in school. Is this evidence for genetic influences on personality and behavior or for some type of psychic connection between these women? Similarities such as these are sometimes attributed to such causes, but these similarities are due to pure chance. How do I know? Because these two women are not twins. They were 1 of 25 randomly created pairs of college students in a study conducted by Wyatt, Posey, Welker, and Seamonds (1984). Wyatt et al. performed the study to determine if the types of amazing identical-twin similarities often reported as evidence of paranormal phenomena could also be found in randomly paired, unrelated people. Their participants were (a) 13 pairs of identical twins who were in high school or college and (b) the 25 randomly paired college students mentioned above (members of the pairs were of the same age and sex). All participants were given a paper-and-pencil personality test and a measure on which they indicated how alike or different they thought they were on a number of attributes, including political beliefs, musical interests, religious preferences, favorite foods, and hobbies. They found that, although the identical twins were more alike than were randomly paired students—a result that could be due to a number of factors, including genetic or environmental similarities—many similarities appeared among the randomly paired students, including the rather "eerie" similarities of the pair mentioned above. Myers (2001) gave the following example of another pair of women with some interesting similarities:

> Patricia Kern of Colorado was born March 13, 1941, and named Patricia Ann Campbell. Patricia DiBiasi of Oregon also was born March 13, 1941, and named Patricia Ann Campbell. Both had fathers named Robert, worked as bookkeepers, and had children ages 21 and 19. Both studied cosmetology, enjoyed oil painting as a hobby, and married military men, within 11 days of each other. They are not genetically related. (p. 95)

Wyatt et al. (1984) concluded, "Our data strongly suggest that it is relatively easy for almost any two people of roughly the same age and sex to find similarities that may seem unusual" (p. 64). This possibility makes sense when you consider the vast number of attributes on which you could compare two people. Although it might be highly unlikely that any two people would match on any *particular* attribute (e.g., green eyes or birthday), it is quite likely that they would match on *some* of the many attributes (e.g., religious affiliation, favorite foods, first names, occupation, hobbies, hair color, number of letters in their last names) one could consider (cf. Gardner, 1972; Wyatt et al., 1984).

Another example of a coincidence that is sometimes attributed to psychic powers is the predictive dream. For example, you might have dreamed that someone you know got into a car accident or became pregnant only to find out later that your dream came true. Paulos (1988) explained this phenomenon as follows:

Assume the probability to be one out of 10,000 that a particular dream matches in a few vivid details some sequence of events in real life. This is a pretty unlikely occurrence, and means that the chances of a nonpredictive dream are an overwhelming 9,999 out of 10,000. Also assume that whether or not a dream matches experiences one day is independent of whether or not some other dream matches experience some other day. Thus, the probability of having two successive nonmatching dreams is, by the multiplication principle for probability, the product of 9,999/10,000 and 9,999/10,000. Likewise, the probability of having N straight nights of nonmatching dreams is $(9,999/10,000)^N$; for a year's worth of nonmatching or nonpredictive dreams, the probability is $(9,999/10,000)^{365}$. Since $(9,999/10,000)^{365}$ is about .964, we can conclude that about 96.4 percent of the people who dream every night will have only nonmatching dreams during a one-year span. But that means that about 3.6 percent of the people who dream every night will have a predictive dream. 3.6 percent is not such a small fraction; it translates into millions of apparently precognitive dreams every year. . . . There's no need to invoke any special parapsychological abilities; the ordinariness of apparently predictive dreams does not need any explaining. What would need explaining would be the nonoccurrence of such dreams. (p. 55)

All of these examples of surprising chance events—from amazing stock predictions to predictive dreams—may *seem* to us to be the result of something other than chance. But, before concluding that such events are the result of paranormal phenomena, fate, conspiracy, or some other cause, ask yourself what role chance might have played in producing the event.

80% Certain, but 0% Sure

We have discussed a number of situations in which people's probability estimates may be biased. Perhaps a more interesting phenomenon is when people estimate probabilities correctly but use that information in an irrational manner. Consider the following hypothetical legal case that Wells (1992, Experiment 3) presented to practicing trial judges, graduate students enrolled in a business statistics course, and undergraduate psychology students:

Mrs. Prob is suing the Blue Bus Company for having caused the death of her dog. At trial, the following evidence was given:

Mrs. Prob testified that she was walking her dog on county road #37 when she heard a large vehicle behind her. She turned around and saw a bus swerving recklessly down the road. She jumped out of the way but the bus swerved and hit her dog, killing him instantly. The incident occurred at 11:40 A.M. The bus continued at a high speed down the road. Unfortunately, Mrs. Prob is color blind and thus does not know the color of the bus.

A county transportation official took the stand, was sworn as a witness, and testified that there are only two bus companies that travel in the county: the Blue Bus Company and the Grey Bus Company. Each company uses the road to run empty buses back to their station after dropping off their passengers. Therefore, one of these two bus companies had to be responsible for the death of Mrs. Prob's dog.

A second county transportation official took the stand and reported that he examined the dead dog and took prints of the tire tracks. These prints were then transferred onto paper and compared to all 10 of the 10 buses owned by the Blue Bus Company and the 10 owned by the Grey Bus Company. The tracks matched 80% of the Blue Bus Company's buses and matched only 20% of the Grey Bus Company's buses. (pp. 741, 743)

In a civil case such as this, a juror should rule for the plaintiff (Mrs. Prob) if the claim is more likely to be true than not true (this is often interpreted as a greater-than-50% probability the claim is true). Most participants in this study correctly estimated that the probability was 80% that a Blue Bus Company bus caused the death of Mrs. Prob's dog, thus their probability estimates were not biased. However, the majority (over 80%) of the participants chose not to rule against the Blue Bus Company and thereby force them to pay damages. Why would these participants make such a decision? Their decision seems irrational in light of their accurate probability judgments and the "more-likely-than-not" rule.

It is interesting that judges and psychology students in another study performed by Wells (1992, Experiment 4) made a very different decision when they were given a slightly different version of the same case. In the new version, called the tire-tracks-belief version, the transportation official did not state that the tire tracks matched 80% of the Blue Bus Company's buses and 20% of the Grey Bus Company's buses, as in the original version, called the tire-tracks version. Instead, he compared the tracks to all of the Blue and Grey buses and "testified that the technique used for matching is correct 80% of the time and, based on this technique, he believed that the bus that ran over Mrs. Prob's dog was a Blue Bus Company bus" (p. 747). Wells' participants were randomly assigned to read one of three versions of this case, two of which were the tire-tracks and tire-tracks-belief versions. He found that the version did not make a difference in participants' subjective probabilities that a Blue Bus Company bus was at fault; regardless of the version they read, the participants judged this probability to be quite high. However, the majority of participants who read the tire-tracks-belief version indicated that the Blue Bus Company should be held liable for the damage, but the majority of the participants who read the original tire-tracks version reached the opposite verdict.

There appears to be a psychological difference between saying that there is an 80% chance that a Blue bus did it (tire-tracks version) and saying that a Blue bus did it based on evidence that is 80% reliable (tire-tracks-belief version). This phenomenon has been called the **Wells effect** (Niedermeier, Kerr, & Messe, 1999). Niedermeier, Kerr, and Messe explained that people are reluctant to rule against the Blue Bus Company after reading the tire-tracks version of the case because they can easily imagine (or mentally simulate) how the Blue Bus Company might not be guilty (they called this the "ease-of-simulation hypothesis"). In other words, if the tire tracks matched 8 out of 10 Blue Bus Company buses and 2 out of 10 Grey Bus Company buses, there are 2 Grey Bus Company buses that could have killed the dog. It is more difficult for people who read the tire-tracks-belief version to imagine how a bus owned by the Grey Bus Company could have been responsible because the transportation official did not mention this possibility.

The ease with which we can mentally simulate events also appears to play a role in another psychological phenomenon illustrated by the following example, used in an experiment by Miller, Turnbull, and McFarland (1989, Study 2):

> John S. is a supervisor in a local manufacturing firm. John is responsible for promoting the employees in his department. In the past he has been accused of being against equal rights and opportunities for women. There is 1 male and 9 females in his department who are potential candidates for promotion. John decides to give these employees a written examination to help with his decision. John grades these exams himself and reports that the highest mark was obtained by a man, whom he promotes. (p. 584)

How suspicious would you be that John's grading of the exam was unfair? How suspicious would you be if instead of 1 male and 9 females, there were 10 males and 90 females in John's department (and, again, the highest mark was obtained by a man—who was then promoted)? In both cases, the probability of a man obtaining the highest score by chance is 10%. Nevertheless, Miller et al. found that their participants were less suspicious when there were 10 males and 90 females than when there was only 1 male and 9 females, even though the participants were aware that the probability of a man obtaining the highest score by chance was the same in both cases. Miller et al. explained that this happens because it is easier to imagine (or mentally simulate) how a man could have obtained the highest score by chance when there are 10 males than when there is only 1 male. Thus, the event seems more normal, and less suspicious, when there are 10 males.

Miller et al. (1989) concluded that their participants were being unfair and irrational. It hardly seems fair to base your suspicion of John on the number of ways a man could have obtained the highest score when the probability of this outcome does not change across the 1-male and 10-male versions. The results of the studies by Miller et al. and of the mock juror studies conducted by Wells (1992) and Niedermeier et al. (1999) suggest that the ease with which we can imagine or mentally simulate improbable events may lure us into making irrational conclusions and decisions, even when we have accurately assessed the probability of such events.

Summary

The concept of probability is as fundamental to our everyday lives as it is to the understanding of statistics. Judgments of the probability of events are important for a wide range of everyday decisions and conclusions, including our assessment of risks, explanations for chance occurrences, and judgments of the causes of others' behaviors. For instance, you might decide to drive rather than fly to some destination if you believe that there is high probability of a fatal crash if you fly. However, researchers have found that the mental shortcuts, or heuristics, we use to estimate probabilities can bias our subjective judgments of probabilities. Keeping in mind some basic rules of probability, such as the multiplication rule and conjunction rule, and understanding the situations in which simple heuristics and mental simulations are likely to lure us into violating these rules may help us avoid misjudging or misusing probabilities that are important to our physical, cognitive, and social well-being.

● PRACTICE PROBLEMS

1. Researchers have found that adolescents and adults who are heavy viewers of television, as compared with light viewers, tend to overestimate the frequency of violence in the world and fear being assaulted themselves. Explain how this finding is related to the concepts of probability and heuristics.

2. A prosecuting attorney is working on the wording he will use in his opening statement during a criminal trial. He is trying to decide whether to say, "the defendant left the scene of the crime" or "the defendant left the scene of the crime for fear of being accused of murder" (cf. Tversky & Kahneman, 1982b, p. 98). He wants to choose the version that represents the most probable event. Which one should he choose? Explain.

3. A friend of yours tells you that she was sitting at home the other night thinking about her sister and, all of a sudden, the phone rang. Her sister called her at the exact moment she was thinking about her. Your friend tells you that this event seems extremely improbable, and she believes it is evidence of a psychic connection she has with her sister. How would you respond?

4. You go to a store looking for a new gas grill for your outdoor patio, and you tell a salesperson that you want to buy one that will work properly for at least 10 years. The salesperson shows you a new model and tells you, "I know that there is a 90% chance that this grill will work for 10 years." He continues, "There are seven major components in this grill that allow it to work properly, and each component has gone through extensive quality testing to ensure there is a 90% chance that it will last for 10 years." Was the salesperson correct when he you told you there is a 90% chance that the grill will work for 10 years? Explain.

3

The Law of Large Numbers

● It is easy to see how people could be led to make erroneous inferences in
: their daily lives by failing to appreciate the stability of population estimates
: based on large samples and the unreliability of estimates based on small
: samples.
:
: *Nisbett and Ross (1980)*
:
:
:
:

● It is time to revisit Question 2 from the questionnaire in Chapter 1.

   2. You are thinking of buying a new computer but cannot decide which one to buy. . . . You visit your college's computer center and get some advice from the person in charge of buying their computers. She tells you that the college purchased computers several years ago. They bought Brand A computers to fill one of their large computer labs, and Brand B computers to fill a second large computer lab. She also informs you that both brands have performed fairly well, but the Brand A computers have had somewhat fewer problems. While discussing computers with your friends, you learn that a good friend of yours bought a Brand A computer a couple of years ago and had nothing but trouble with it. Only two months after purchasing his computer, the hard drive failed. After the hard drive was replaced, the monitor died. Once that was replaced, the memory chips failed. After talking with your other friends, you discover that one of them purchased a Brand B computer a couple of years ago and she absolutely loves it. Her computer has performed flawlessly and she highly recommends you buy one. Which brand (A or B) would you choose?

When you first read this question, did you choose Brand B? Perhaps your friend's horrible experience with a Brand A affected your decision. Such a disastrous experience may be a very memorable and highly "available" event that could have a substantial influence on your estimate of the reliability of Brand A computers. You may recall from Chapter 2 that our use of the availability heuristic could lead to biased judgments in such situations.

The quote at the beginning of this chapter suggests a useful statistical perspective from which to view the computer-choice question. In an ideal world, you could examine the failure rates of *all* Brand A and Brand B computers (i.e., the population failure rates) and use them as an estimate of the future reliability of each brand (assuming failure rates are stable over time). But you do not have this information, so you must estimate the population failure rates. As dictated by the **law of large numbers,** a larger sample will provide a better estimate of the population than will a smaller sample. This implies that the information you received from the woman who purchases computers for your college is much more valuable than is the information you received from your friends. Your friends' experiences suggest that Brand B is better than Brand A, but a sample of $n = 1$ is practically useless for drawing any conclusion about the population failure rates of each brand. The information from the woman who works for the college reflects much larger sample sizes (i.e., the large computer labs should have a sizable number of computers) and thus should provide better estimates of the computers' population failure rates. Because the information she gave you suggests that Brand A is better than Brand B, you should choose Brand A.

The law of large numbers (LLN) is an elegantly simple yet immensely important rule. You might be surprised by the variety of situations to which it applies. Paulos (1988) called the law of large numbers "one of the most significant though often misunderstood theorems in probability theory" (p. 116). This chapter will examine some of the everyday situations to which this rule applies. We will first take a closer look at the LLN itself. Afterward, we will examine how good we are at applying the LLN when we draw conclusions about everyday events and discuss how we sometimes violate this law by (a) relying too heavily on single-case evidence and (b) drawing inaccurate conclusions about random events.

What Is the Law of Large Numbers?

Both laypeople and scientists often have to rely on samples to estimate population values because we do not have access to the entire population in which we are interested. For example, if you wanted to estimate the percentages of male and female students at your college, you would probably not be able to count every single student to determine them. Instead, you would use a sample of students as a basis for estimating the population. Assume that 65% of the students at your college are women and 35% are men (these are the population percentages you wish to determine). Also assume that I have a list of all of the students at your college, and I offer you the opportunity to select a random sample of students from this list. You will use this sample to estimate the percentages of men and women at your college. You can choose a sample of either 5 students or 50 students. Which sample size would you choose? It should be obvious that a sample size of 50 is a better choice because it will give you a more accurate estimate of the population. If you chose a sample of 5 students, it is possible that all of them would be men; this would seriously distort your estimate of the percentage of men on campus. With a sample of 50 students, it is highly improbable that all of them would be men. This illustrates one way of conceptualizing the LLN: Large samples provide us with better estimates of populations than do small samples.

As mentioned in the chapter-opening quote, one implication of the LLN is that small samples provide us with population estimates that are much less stable or reliable than those based on large samples. In other words, if I had 10 students each draw a random sample of 5 students from my list, the 10 estimates of the percentage of males on campus might be very different from each other. One student might obtain a sample with 80% men, another might have 40% men, a third might have 0% men, and so on. However, if 10 students each drew a random sample of 50 students, the 10 estimates would resemble each other more closely. In general, if we were to take all possible random samples of size 5 from a population, we would obtain sample values (or statistics) that are much more spread out around the population value (or parameter) we are estimating than if we took all possible random samples of size 50. The greater variability of sample values from small samples, as compared with large samples, is represented graphically in Figure 3.1. This figure shows two hypothetical **sampling distributions**—that is, distributions of sample values from all possible random samples of a certain size—and you can see that the sample values from large samples are more tightly clustered around the population value than are the values from small samples.

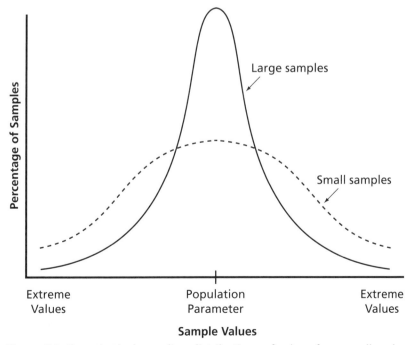

Figure 3.1 Hypothetical sampling distributions of values from small and large samples

To consider another way of looking at the LLN, think about flipping a coin. If you have a fair coin, the probability of obtaining heads on a flip is 50%. If you were to flip the coin three times, it is easy to imagine obtaining heads on all three flips, even though the probability of heads on each flip is only 50%. However, if you were to flip the coin 1,000 times, it is very difficult to imagine that you would obtain heads on all of the flips. Instead, you would expect to obtain heads on roughly 500—or 50%—of those flips. Thus, in the long run (that is, with a larger sample), the difference between the relative frequency of an event (e.g., proportion of heads in a series of coin flips) and the probability of the event (e.g., probability of heads = 50%) will approach zero. This is the law of large of numbers in operation.

How Good Are We at Applying the LLN?

The Representativeness Heuristic and Failures to Apply the LLN

Consider the following questions posed to participants in studies performed by Kahneman and Tversky (1972):

> 1. A certain town is served by two hospitals. In the larger hospital about 45 babies are born each day, and in the smaller hospital, about 15 babies are born each day. As you know, about 50% of all babies are boys. The exact percentage of baby boys, however, varies from day to day. Sometimes it may be higher than 50%, sometimes lower. For a period of 1 year, each hospital recorded the days on which more than 60% of the babies born were boys. Which hospital do you think recorded more such days? (p. 443)

> 2. A medical survey is being held to study some factors pertaining to coronary diseases. Two teams are collecting data. One checks three men a day, and the other checks one man a day. These men are chosen randomly from the population. Each man's height is measured during the checkup. The average height of adult males is 5 ft 10 in, and there are as many men whose height is above average as there are men whose height is below average. The team checking three men a day ranks them with respect to their height and counts the days on which the height of the middle man is more than 5 ft 11 in. The other team merely counts the days on which the man they checked was taller than 5 ft 11 in. Which team do you think counted more such days? (pp. 443–444)

If your reasoning was consistent with the law of large numbers, you should have chosen the smaller hospital in the first question and the team that checked only one man a day in the second question. In both cases the sample size is smaller and is therefore more likely to produce extreme values—that is, results that deviate from the population values of 50% in the first problem and 5 ft 10 in in the second problem. However, Kahneman and Tversky (1972) suspected that people would answer these questions incorrectly because they would rely on the representativeness heuristic rather than the LLN to determine their answers. In other words, they suspected that people would judge the probability of obtaining various sample results according to the degree of *similarity* between the sample results (or statistics) and the population parameter. Because small and large samples of more than 60% boy babies are equally similar to the population value of 50% (i.e., they have the same statistic), Kahneman and Tversky predicted that their participants would ignore the sample sizes and judge these sample outcomes as equally likely. This result would suggest that people rely on the representativeness heuristic rather than the LLN to answer this question. For the same reason, they predicted that participants would say the two medical-survey teams should have an equal number of days on which they recorded heights greater than 5 ft 11 in. Consistent with their predictions, they found that the majority of their participants said that (a) the larger and smaller hospitals recorded "about the same" number of days on which more than 60% of the babies born were boys and (b) the two teams recorded "about the same" number of days on which the man they checked was taller than 5 ft 11 in. According to Kahneman and Tversky, sample size did not influence participants' judgments because it was not representative of, or similar to, any characteristic of the population.

Kahneman and Tversky (1972) also discovered that the representativeness heuristic seems to guide people's intuitive notions about sampling distributions. In one study, they had Israeli participants imagine that a center recorded the average height of a certain number of men—10, 100, or 1,000—every day. They told participants that the average height of the Israeli male population was between 170 and 175 cm, and they asked participants to estimate the probability that the samples would have an average height that fell into various categories (e.g., 165–170 cm, 170–175 cm, 175–180 cm). Of course, the law of large numbers dictates that larger samples will provide more accurate estimates of the population mean and, therefore, the sampling distributions of larger samples will contain a much larger percentage of sample means in the 170 to 175 cm range than will sampling distributions based on smaller samples (see Figure 3.1). In other words, sampling distributions created from larger samples have much less variability than do sampling distributions created from smaller samples. However, as they predicted, Kahneman and Tversky found that sample size had no impact on the participants' subjective sampling distributions. Participants who were asked to generate sampling distributions for small sample sizes produced distributions identical to those generated by participants given larger sample sizes. An average height of less than 165 cm (for example) in a small sample is just as similar to the population mean (170–175 cm) as are large samples with an average height less than 165 cm in a large sample. Because they are equally similar, people judge them to be equally likely. However, as we have seen, extreme values are much more likely to be found in small samples.

Do People Have an Intuitive Version of the LLN?

Paulos (1988) said that the LLN is "often misunderstood," and Kahneman and Tversky (1972) stated, "The notion that sampling variance decreases in proportion to sample size is apparently not part of man's repertoire of intuitions" (p. 444). However,

Fong, Krantz, and Nisbett (1986) claimed, "People possess an abstract inferential rule system that is an intuitive version of the law of large numbers" (p. 253). Who is correct? Do people apply the LLN in everyday situations that require statistical reasoning? Although Kahneman and Tversky's experiments suggest the answer to this question is "no," other researchers suggest the answer is "yes and no." For some types of problems, people are quite good at reasoning in line with the LLN; but for other types of problems, their reasoning often violates the LLN. To give you a sense of what distinguishes these two types of problems, consider the following examples used by Fong et al.:

> 1. At Stanbrook University, the Housing Office determines which of the 10,000 students enrolled will be allowed to live on campus the following year. At Stanbrook, the dormitory facilities are excellent, so there is always great demand for on-campus housing. Unfortunately, there are only enough on-campus spaces for 5,000 students. The Housing Office determines who will get to live on campus by having a Housing Draw every year: Every student picks a number out of a box over a 3-day period. These numbers range from 1 to 10,000. If the number is 5,000 or under, the student gets to live on campus. If the number is over 5,000, the student will not be able to live on campus.
>
> On the first day of the draw, Joe talks to five people who have picked a number. Of these, four people got low numbers. Because of this, Joe suspects that the numbers in the box were not properly mixed and that the early numbers are more favorable. He rushes over to the Housing Draw and picks a number. He gets a low number. He later talks to four people who drew their numbers on the second or third day of the draw. Three got high numbers. Joe says to himself, "I'm glad that I picked when I did, because it looks like I was right that the numbers were not properly mixed." What do you think of Joe's reasoning? (pp. 282–283)
>
> 2. Gerald M. had a 3-year-old son, Timmy. He told a friend: "You know, I've never been much for sports, and I think Timmy will turn out the same. A couple of weeks ago, an older neighbor boy was tossing a ball to him, and he could catch it and throw it all right, but he just didn't seem interested in it. Then the other day, some kids his age were kicking a little soccer ball around. Timmy could do it as well as the others, but he lost interest very quickly and started playing with some toy cars while the other kids went on kicking the ball around for another 20 or 30 minutes." Do you agree with Gerald's reasoning that Timmy is likely not to care much for sports? (p. 287)

In the first problem, Joe based his conclusion on very small samples of all the students who picked numbers each day. Because small samples provide very unreliable estimates of population values, the difference he observed in the proportion of low numbers in the first and second sample of students could have been due to chance. Even if exactly half of the numbers drawn each day were low numbers—that is, if the population proportion of low numbers selected each day was .50—it remains possible that the proportion Joe found would be .80 for a sample of 5 people one day and .25 for a sample of 4 people on another day. Thus, Joe should not conclude that the numbers were not properly mixed.

Of course, you probably answered this question correctly before you even read the explanation. Fong et al. (1986) found that the majority of their participants solved these types of problems accurately. They called this type of problem "probabilistic," because it is clear that (a) Joe polled only a small sample of all the students who picked numbers each day and (b) chance played a role in producing the outcome. However, the role of sampling and chance are less easily recognized in the second problem, and that is why Fong et al. called this second type of problem "subjective." Fong et al. found that the majority of their participants failed to apply the LLN to subjective problems. Were you able to apply the LLN to this problem? From a statistical perspective, it is not very different from the first problem. Gerald is trying to estimate a population value (i.e., Timmy's interest in sports across various sports, times, and places) based on a very small sample (i.e., Timmy's interest in playing catch with a neighbor one day and his interest in kicking a soccer ball with some kids on another day). Timmy could be interested in sports, but we would not expect him to demonstrate that interest at every moment in every situation. Thus, if we were to make two observations of his sports-related behavior in any given month, it is certainly possible that we would find him showing disinterest each time. Moreover, given the unreliability of small samples, if a second person also made two observations of Timmy's sports-related behavior in the same month (on days different from those observed by Gerald), she might come to a very different conclusion about Timmy's interest in sports.

Although Fong et al. (1986) found that people are generally likely to apply the LLN to probabilistic problems like the Stanbrook University example, real-life failures to apply the LLN in similar situations have been documented. For example, in the early 1970s, the U.S. Supreme Court decided that, for some types of cases, courts could use 6-person juries rather than the usual 12-person juries. They reasoned that the reduction in jury size would have a negligible impact on the degree to which the jury represented a community's diversity. Do you agree? Imagine that you are an African-American male who has accused a Caucasian police officer of brutally beating you after pulling you over for a routine traffic violation. If you believed that the race of the jurors would affect their judgments of this case, would you mind if the case appeared before a 6-person jury instead of a 12-person jury? According to the LLN, a sample of 12 people should be more representative of a population than would a sample of 6 people; thus, a 12-person jury would be more likely to contain an African American. Certainly, a community's diversity is more likely to be represented in a 12-person jury. Unfortunately, the Supreme Court's decision was based on reasoning that violated the LLN, even though jury selection clearly involves sampling and chance factors and therefore is similar to what Fong et al. called a "probabilistic" problem.

The Impact of Single Cases

Failing to understand that a sample of 12 citizens will be more representative of a community's diversity than will a sample of 6 citizens is certainly a violation of the LLN. However, it is not quite as blatant a violation as is our tendency to be influenced by single-case evidence. We sometimes form ill-advised, sweeping generalizations from single cases. Marketers seem to be very aware of our affinity for single-case evidence. Magazine advertisements and television commercials are filled with testimonials given by individuals proclaiming the effectiveness of pain killers, weight-loss programs, herbal remedies, health clubs, and other products. For instance, a recent commercial for a chain of fitness centers in Cincinnati featured a woman who appeared in a similar commercial for this chain many years ago. The announcer in the recent commercial explained that, because this woman exercised at their fitness centers, she is still thin and beautiful—even though many years have passed and she has given birth to children. The implication, of course, was that everyone could achieve similar results by joining their fitness center, but this is an awfully risky generalization to make based on a single case.

Another recent commercial—which was a rather bold affront to scientists and others who are adept at statistical reasoning—featured a man selling a pain killer who said something similar to "Sure, I know that scientific studies have proven that this brand is effective at relieving pain, but I don't care about that; I've tried it and I know it works." In effect, the commercial suggested that you should not rely on an adequate-sized sample (typically used in scientific studies) when you can have a sample of $n = 1$!

Of course, single-case evidence is suspect not only because of the small sample size, but also because of the numerous possible explanations for any observed effect. For example, a woman's thin physique might be due to any number of causes other than her visits to a fitness center. Perhaps she has a naturally high metabolic rate, a job that requires a lot of physical exertion, a low-calorie diet, or some combination of these factors. Similarly, if you take a pain killer and notice improvement, that improvement could have been caused by the passage of time, your belief that the pain killer would work (a **placebo effect**), or some other factor that happened to coincide with your ingestion of the drug. Therefore, we need to be extremely cautious about drawing conclusions from single-case evidence.

Single Cases Versus Group Statistics

What is even more surprising than the degree to which we are influenced by single cases is that sometimes they have a larger impact on us than do group statistics based on larger samples. For example, imagine you are trying to decide whether to take a psychology course titled "Social Influence" or one named "Social Perception" next semester. In addition to the descriptions of each course, you have two other pieces of information about them: (a) the average course ratings from the students who took these courses last year and (b) the opinions of a psychology major who took both courses last year. Which of these two pieces of information would you be most likely to rely on for choosing a course? Borgida and Nisbett (1977) investigated this issue with prospective psychology majors enrolled in an introductory psychology class at the University of Michigan. They told the students that the psychology department was trying to determine which psychology courses students were likely to take over the next several years. To help the students make their decisions, they were shown descriptions of some of the courses. One group of the students was also shown course ratings (labeled "excellent," "very good," "good," "fair," or "poor") for the courses, and they were told the ratings were averages based on students—ranging in number from 26 to 132—who evaluated the courses the previous semester. Instead of the average course ratings, a second group of students received face-to-face comments about the courses from between one and four psychology majors who had taken the courses. These psychology majors gave each course an overall rating (e.g., "excellent," "very good," "good,"), and the averages of these ratings were identical to the average course ratings given to the first group of students. A third group of students was given no course evaluations at all. Then, the students in each group were asked to indicate which psychology courses they were likely to take.

Borgida and Nisbett (1977) found that the students who received the average course ratings (or group statistics) were no more likely to take courses with favorable evaluations or avoid courses with unfavorable evaluations than were the other two groups of students. Thus, the group statistics appeared to have little impact on their course choices. However, the students who received face-to-face evaluations from a very small number of psychology majors were influenced by those evaluations. As compared with the other groups of students, their preferences were stronger for courses given a favorable review and weaker for courses given an unfavorable review.

Borgida and Nisbett's findings are just the opposite of what we would expect if the students' decisions were governed by the law of large numbers. They speculated that this result was at least partly due to "a lack of understanding of the fundamental principles of statistical inference" (p. 269). As they explained:

> In the present research, the opinion of a single, perhaps highly atypical individual is probably taken as quite indicative of opinions in general. The similar opinion of a second, let alone a third or fourth individual, is then regarded as more than sufficient to clinch the matter. In the face of such faith in the reliability of small samples, the addition of dozens of further opinions would be regarded as sheer redundancy. The value of small sample information is thus inflated and the value of large sample information is degraded. (p. 270)

"Person-who" Statistics

Another disturbing illustration of our preference for single-case evidence over group statistics is the use of "person-who" statistics (cf. Nisbett & Ross, 1980; Stanovich, 2001). **Person-who statistics** involve using a single case to contradict the implications of group statistics. This is perhaps the most blatant violation of the law of large numbers one could make. Here are some examples of the types of arguments you might hear from a person fond of person-who statistics:

1. "Smoking supposedly causes lung cancer, but I know a *person who* smoked a couple of packs of cigarettes a day and lived to the age of 90 with no sign of cancer."
2. "The information in *Consumer Reports* magazine suggests that Honda automobiles are very reliable, but I know a *person who* owned one and had nothing but trouble with it."
3. "The chair of the psychology department told me Professor Haines gets excellent ratings from students, but I know a *person who* had her for a class and had a dreadful experience."

Several years ago, a faculty member at my college tried to use a person-who argument to prove to me that there was a flaw in a proposed new form for student ratings of teacher performance. At the time, I was the chair of a committee that was developing that new form. The committee members had carefully reviewed the scientific research on the reliability and validity of student ratings and we tried to include items on the form that were valid indicators of teacher performance. For example, one item required students to rate an instructor on the degree to which he or she was organized in class. This item was included because previous research indicated that instructor organization is related to student achievement (e.g., Marsh, 1984). Nevertheless, the faculty member did not believe that organization was an important quality of a good teacher. To prove her point, she said "I once had a *college professor who* was very disorganized, and I learned a lot from him." The implication of her argument was that her experience with one professor somehow invalidated the proven relationship (based on group statistics) between instructor organization and student achievement.

At some level, person-who arguments seem to have an intuitive appeal. However, single-case evidence simply cannot refute group statistics. Perhaps part of the problem is that people fail to understand that probabilistic statements do not apply to every single case (cf. Stanovich, 2001). In other words, smoking cigarettes increases the *probability* that a person will get lung cancer. However, that does not mean that every single person who smokes will develop lung cancer; and it does not mean that finding a single cancer-free smoker negates the probabilistic relationship between smoking and lung cancer. Imagine a medical researcher who injects thousands of people with an influenza virus, and most of them develop the flu within a week. If I were to show you one person from that group who did not develop the flu, would you conclude that this virus does not cause the flu? Of course not! You would understand that being exposed to the virus increases the probability of developing the flu, but it does not guarantee such a result. You should apply similar reasoning the next time you hear someone use person-who statistics in an attempt to refute group statistics.

The behavior of contestants on the popular television game show *Who Wants to Be a Millionaire?* suggests that people are able to ignore single cases in favor of group statistics in some situations. Imagine you are a contestant on this show and you are asked the following $32,000 question:

The Euphrates River flows through which of the following countries?

a. Iraq   b. Iran
c. Egypt   d. Afghanistan

You are about 70% sure that the answer is Iran, but you want to be absolutely certain because this question is worth a lot of money. So, you decide to use one of your "life lines": You ask the audience. The most frequent answer selected by the audience is Iraq. What should you do, go with the audience's choice or your original choice? It seems that most contestants elect to go with the audience in this situation. Thus, they are ignoring the single case (their original answer) in favor of a group statistic (the most frequent answer selected by the audience). This decision is consistent with the law of large numbers.

We could view the audience and the contestant as samples of different sizes from a larger population of people. If we assume that the majority of the people in this population know the answer to the $32,000 question, then the modal (i.e., most frequent) response of a large audience is more likely to be correct than is the response of one contestant. Although many contestants elect to go with the audience's modal answer, it is unclear whether this behavior indicates an understanding of the implications of the LLN in this situation. They might choose that strategy because of host's frequent reminders that the audience is almost always correct, even if they do not understand *why* the audience is almost always correct. (By the way, the answer to the question is Iraq.)

Perceptions of Randomness

The other day I asked a student in my class to flip a quarter and tell the class the result. She said "tails." I asked her to do it again, and she said "tails." I asked her to flip it a third time, and once again she said "tails." I told her to hold on to her quarter for a minute, then I asked the whole class to predict what the outcome would be if I requested a fourth toss of the coin. What would you predict? Almost all of my students predicted the result would be heads. Of course, if the coin is a "fair" coin, it is equally likely to land on heads or tails on the fourth toss. If the students understand that the probability of either outcome (heads or tails) is 50%, why would they all predict heads? A number of them should predict tails. This is an example of a phenomenon called the **gambler's fallacy**—the belief that a long run of a particular outcome of a chance event, such as a coin landing on tails or a roulette wheel landing on red, makes a different outcome more likely on the next try. However, this is an erroneous belief because the outcome of a single coin flip does not depend on previous outcomes. Even if you flipped a coin 20 times and it landed on tails each time, it is equally likely to land on tails or heads on the next flip.

Ironically, people's understanding that a coin should land on heads, or tails, about 50% of the time may be partly responsible for their tendency to predict heads for the next toss after a long run of tails. In other words, because people know that the probability of either outcome is 50%, they believe that about half of a series of coin tosses should result in heads and half should result in tails. Thus after a run of tails, it seems that heads are "needed" to make the series of coin tosses look representative of a fair coin. Certainly, the sequence TTTH seems more representative of a fair coin than does the sequence TTTT; however, the two sequences are equally likely (probability = $.5^4$, or 6%, according to the multiplication rule).

According to the law of large numbers, we should expect the proportion of heads in a series of coin flips to be close to .50 only after a very large number of flips. The proportion of heads in a small series of coin flips is more likely to deviate from .50 because smaller samples are less representative of populations than are larger samples. However, people appear to believe that the law of large numbers applies to small numbers (or samples) as well. In other words, people believe that even small samples should provide stable, accurate estimates of population values. Tversky and Kahneman (1971) called this a belief in the *law of small numbers*. They suggested that this belief is a result of the use of the representativeness heuristic, as they explained:

> We submit that people view a sample randomly drawn from a population as highly representative, that is, similar to the population in all essential characteristics. Consequently, they expect any two samples drawn from a particular population to be more similar to one another and to the population than sampling theory predicts, at least for small samples. (p. 105)

Thus, one implication of using the representativeness heuristic to judge random sequences is that people expect the sequences to be self-correcting (e.g., a run of 3 tails should be corrected by heads on the upcoming tosses). A second implication is that people expect fewer runs in random sequences than should occur by chance. Considering coin tosses, for example, the sequence HTTHTHHT might be regarded as more likely than the sequence HHHHTTTT because the former sequence is more similar to our notion of a random process, but the two sequences are equally likely to occur. This misconception could lead us to conclude that random patterns are not random at all because we see clusters (or runs) of outcomes that appear nonrandom. For instance, we may think that a gambler has a special talent for tossing dice—or that he has a "hot hand"—if he tosses a seven several times in a row. As another example, the news media in Cincinnati recently focused on several serious highway accidents that occurred within a period of several months near the intersection of two major highways. The seemingly unavoidable conclusion was that this is a particularly dangerous area in which to drive. However, one would expect even a random pattern of highway accidents to contain some clusters in specific areas. Although the clusters occurred by chance, people might be tempted to conclude that something in those areas—such as a high speed limit or flaw in the design of the highway—is causing an inordinate number of accidents. Similarly, Paulos (1994) stated:

> Determining, for example, whether clusters of a particular disease constitute evidence of something seriously awry—or merely a coincidental clumping—is not easy. You may notice, for instance, that a lot of people in your neighborhood seem to be getting brain cancer. But random distributions are not homogeneous. That is, a perfectly even distribution of brain cancer victims across all 50 states would be highly unlikely—far more unlikely than the chance concentrations that occur here and there. (pp. 34–35)

Randomness and Belief in the "Hot Hand"

Imagine you are a basketball player who is playing in an important playoff game. Your team is down by only two points, but there are only 10 seconds left in the game. You have the ball at mid-court, and you are about to pass the ball to one of two teammates who are near the basket. Both teammates are good players, but one of them made her last four shots in this game. The other teammate missed her last four shots. Which teammate should you pass it to? The answer is that it does not matter which player you pass it to (assuming they are equally close to the basket and equally well guarded by the opposing team). Are you surprised? Did you think you should pass it to the player who had the hot hand (i.e., the one who made her last four shots)?

As implied in the earlier example of a gambler who rolls several sevens, the **hot hand belief** is related to the tendency to perceive clusters of events as nonrandom. The expectation that random sequences should contain fewer runs, or clusters, than they typically do may underlie people's belief in the hot hand or streak shooting. A belief in the hot hand is exemplified by the certainty expressed by basketball players, coaches, and fans that basketball players have hot streaks in which they are more likely than normal to make shots (Gilovich, 1991; Gilovich, Vallone, & Tversky, 1985). For example, Gilovich, Vallone, and Tversky found that 91% of the college basketball fans they surveyed agreed that a player has "a better chance of making a shot after having just *made* his last two or three shots than he does after having just *missed* his last two or three shots" (p. 297). Moreover, 84% of the fans agreed that "it is important to pass the ball to someone who has just made several (two, three, or four) shots in a row" (p. 297).

Of course, basketball players sometimes make a number of shots in a row, but this is not evidence of a hot hand unless such runs are longer or more frequent than would be expected by chance. Gilovich, Vallone, and Tversky (1985) investigated the accuracy of hot hand beliefs by examining the performance of basketball players on the court. In one study, they obtained the field goal records for the home games of the

Philadelphia 76ers during the 1980–1981 season. For nine of the team's players, they calculated the conditional probability of a hit (i.e., making a basket) given (a) one, two, or three hits on previous shots or (b) one, two, or three misses on previous shots. They found that the players made an average of 51% of their shots after a hit on the previous shot, and they made 54% of their shots after a miss on the previous shot. Thus, contrary to the hot hand hypothesis, the probability of a hit was slightly higher after *missing* the previous shot than it was after making the previous shot. They found similar results for the probability of a hit after two or three hits on previous shots as compared with the probability of a hit after two or three misses on previous shots.

You might argue that Gilovich et al. (1985) did not conduct a fair test of the hot hand hypothesis because (a) players might select harder shots after hits and easier shots after misses or (b) defensive players might put more pressure on players who hit several shots in a row, thereby making them more likely to miss their next shot. To eliminate the potentially contaminating effects of shot selection and defensive pressure, Gilovich et al. conducted two additional studies. In one of these studies, they analyzed the free throws attempted by a number of Boston Celtics players during two basketball seasons. Free throws are shot from the same distance each time and are not subject to pressure from defensive players. However, contrary to the hot hand hypothesis, the players were not more likely to hit a free throw after making the previous free throw than they were after missing the previous free throw. In their final study, Gilovich et al. had college basketball players—one player at a time—take 100 shots from different points along an arc drawn on the floor. These shots were all the same distance from the basket, and there were no defensive players. Once again, contrary to the hot hand hypothesis, they found that the players were no more likely to make a shot after hitting the previous shot (or several shots) than they were to make a shot after missing the previous shot (or several shots).

More recently, similar results were obtained with the National Basketball Association (NBA) players who participated in the NBA three-point shooting contests in 1998 and 2000. Similar to Gilovich et al.'s final study, players in these contests took 25 consecutive shots at different points along the three-point shooting line, which maintains an equal distance from the basket. For each player, Alan Reifman (personal communication, February, 2000) calculated the probability of a hit given a hit on the previous shot and the probability of a hit given a miss on the previous shot. Consistent with the findings of Gilovich et al., Reifman found no convincing evidence of a hot hand among the NBA players in the three-point contests.

How could basketball players, coaches, and fans form and maintain a belief in the hot hand if there is no evidence of streak shooting in the performances of the players they watch? Consider the shooting record of Dirk Nowitzki, who participated in the NBA's three-point contest in 2000 (O = miss, X = hit):

OXOOOXOXOOXOXOXXXXXXOXXXX

It is easy to see how people could develop a belief in the "hot hand" after witnessing shooting sequences like this one. The pattern of hits and misses does not match our notion of a typical random pattern. There are a couple of long streaks of hits near the end of the sequence, creating the illusion of streak shooting. In order to determine whether this sequence is evidence of a hot hand, we could, like Gilovich et al. (1985), examine whether the probability of a hit following a hit is greater than the probability of a hit following a miss. If you look at each of the 14 shots Dirk Nowitzki attempted after *making* the previous shot, you will see that he had 8 hits and 6 misses; thus, the probability of a hit given a hit on the previous shot was 57% (8/14). If you look at each of the 10 shots he attempted after *missing* the previous shot, you will see that he had 7 hits and 3 misses; thus, the probability of a hit given a miss on the previous shot was 70% (7/10). Once again, the hot hand hypothesis is not supported by the data. Basketball players do have streaks during which they make a number of consecutive shots, but these streaks are generally no longer or more frequent than would be expected by chance.

A belief in the hot hand among basketball players seems harmless enough, but it could become very costly in situations where a player decides to pass the ball to a teammate who made his last few shots and overlook a teammate who missed his last few shots. If the "cold" teammate is less heavily guarded but just as skilled as the "hot" teammate—or if the cold teammate has a higher average scoring percentage than the hot teammate—passing the ball to the hot teammate could decrease the probability of scoring and possibly result in losing an important game.

Summary

The law of large numbers is a simple but important rule that applies to a wide variety of everyday situations. Simply stated, the law of large numbers dictates that large samples provide us with better estimates of populations than do small samples. Researchers have discovered that people are often able to reason in a manner consistent with the law of large numbers in situations where the roles of sampling and chance are clear. However, we may fail to apply the law of large numbers in some situations, such as when we are confronted with compelling single-case evidence, random sequences with long runs of a particular outcome, or subjective problems in which the roles of sampling and chance are difficult to detect. Failing to appreciate the implications of this simple rule may create a diversity of problems, such as choosing an inferior consumer product, making a costly gambling decision, misperceiving random events as signs of danger, or losing an important basketball game because you passed the ball to the wrong player.

● PRACTICE PROBLEMS

1. A new restaurant, named Sunset Café, opens in your town. You visit it one day and absolutely love their food. You tell a good friend about your experience, and she informs you that she went there during the same week you

visited. However, her reaction was much different; she thought the food was mediocre at best. Because you and your friend have very similar taste in food, you are puzzled by this disagreement. Why would this happen?

2. The psychology faculty at a small college are about to hire a new psychology faculty member. They invite three candidates to campus for interviews. As part of the interview process, each candidate must teach a class for 30 minutes while the psychology faculty observe. The first candidate teaches the class quite well; she lectures in an engaging manner and gets the students actively involved by using a brief exercise. The second candidate's performance was good, but her lecture was less engaging than the first candidate's lecture. The third candidate's performance was the worst; unlike the other two candidates, she did not get the students actively involved in class, and her lecture was no better than that of the second candidate. The psychology faculty also have additional information about the third candidate. She taught two classes at this college while she was in graduate school, and the psychology faculty have the student ratings from these classes. The student ratings for each of her classes are excellent. Which candidate should the psychology faculty choose? Why?

3. What follows is a true story. In the spring of her senior year in high school, a young woman went to an assessment/registration session at the college she planned to attend that fall. As part of its assessment process, the college had the woman—as well as all other incoming students—write a very brief (about two paragraphs long) essay about her experiences in high school. Two faculty members who taught writing classes at the college judged the essay. They evaluated the grammar, spelling, punctuation, and clarity of the essay. They judged the woman's essay to be of poor quality, and they recommended this woman be placed in a remedial English course that focused on the development of basic writing skills. The incoming students who passed this writing test, on the other hand, were placed in a somewhat more advanced writing course—a course that all students were required to take before graduating.

After receiving the recommendation from the writing assessment team, the faculty member who was advising the woman noticed that she had performed very well in the three English classes she took in high school and on her SAT verbal test. The discrepancy between this information and the writing team's judgment seemed odd to the advisor. He told the woman about the team's recommendation and asked her how she felt about her writing skills. She informed him that she did quite well in high school and received As on the papers she had written. The advisor discussed this information with members of the writing assessment team as well as the college's academic dean. Although the advisor felt strongly that the woman should not have to take the remedial English course, the writing assessment team and the Dean required her to register for the course. Why do you think the advisor felt this way? If this issue had been turned over to you for a final decision, what decision would you make? Why?

4. In a course I taught, we discussed a number of scientific studies that found that kids who watch violent television programs tend to exhibit more violent behaviors than do kids who are not exposed to violent programs. One student commented, "I don't believe that is true. I watched some violent programs when I was a kid and I am not a violent person." Is this a sound argument? Explain.

Estimation of Population Parameters: The Problem of Sample Bias

> One of the most pervasive characteristics of social judgment is that people generalize from themselves to others. They tend to believe that most other people share their own preferences, habits, or sentiments.
>
> *Krueger and Clement (1997)*

The questionnaire in Chapter 1 asked you to consider the following situation:

3. Several years ago, a student of mine came to my office to discuss the poor grade she received on an exam in my class. Her anger was clearly visible as she explained her belief that my exams were too difficult. She had reached this conclusion after talking with a couple of other students in the class who also performed poorly on the exam. If you were this student, would you reach the same conclusion? Would you be angry?

Apparently, this student was inferring that most students in the class had performed poorly on the exam. As illustrated by the discussion of the law of large numbers in Chapter 3, it is not wise to draw firm conclusions about a population (e.g., the exam performance of the entire class) based on a very small sample (e.g., the performance of a couple of students in the class). However, that is not the only problem with the student's conclusion. Another serious problem is the way in which her sample of students was selected. Although I do not know for sure how she selected her sample, I know that the sample was not representative of the entire class, because the grade distribution for the entire class was very good. In fact, most of the students in the class achieved As and Bs on the exam. Perhaps she noticed a couple of students near her who expressed disappointment when I handed back the exams in class, and she decided to discuss her exam results with them. Obviously, this would be a very biased sample of students; they were selected because they did not perform well on the exam. To use this sample to estimate the performance of the entire class would be a violation of a very important, fundamental rule of statistical inference: *If you want to make an accurate estimate of a population value, you should rely on a sample that is representative of the population from which it came.*

Certainly, if a student wanted to make an accurate estimate of the number of students in her class who performed poorly on an exam, polling a couple of students near her would not be the best way to make that estimate—especially if those students were selected because they performed poorly on the exam. The exam scores of those students are not likely to be typical, or representative, of the class as a whole. To arrive at a more accurate estimate of the grades obtained by the entire class (i.e., the population frequency distribution), she might select a more representative sample of students. Better yet, she might bypass sampling altogether and simply obtain the entire grade distribution from her instructor. Of course, it is not always possible to obtain information about the entire population in which we are interested. We often have to base our population estimates on samples from the population. Let's take a closer look at the process of estimation and then consider more real-life examples.

The Process of Estimation

The process of **estimation** involves using sample values (i.e., sample statistics) to estimate population values (i.e., population parameters). For example, a researcher might want to estimate the average income of people in Cincinnati, the average age at which men in the United States get married, or the percentage of registered voters who would vote for a political candidate. In each case, the researcher would want to carefully select a sample of people on which to base her estimate. If she were interested in making inferences about the income of people in Cincinnati, she would want her sample to be representative of those people. If she were interested in making inferences about the population of registered voters, she would want her sample to be representative of that group. Selecting a biased, or unrepresentative, sample from either of these groups would result in inaccurate estimates.

Imagine, for instance, that a researcher wanted to estimate the percentage of students at a small college who drink five or more alcoholic drinks per week. In order to obtain a representative sample, he might select a **random sample** of all the students at the college. For a sample to be truly random, each member of the population must have an equal chance of being included in the sample. Thus, the researcher might obtain a list of all students registered at the college, put each student's name on a small piece of paper, toss all of those names into a box, mix them up, and choose his sample from the box. Of course, researchers nowadays would be more likely to have a computer select the random sample, but the concept of random sampling seems easier to visualize by imagining selecting a sample from well-mixed pieces of paper in a box.

Once the researcher selected his sample, he would survey the students about their drinking habits and use the results to create an estimate of the drinking habits of the population of students at the college. Consider how his estimate might be different if he had selected his sample in a nonrandom, or biased, fashion. For example, if the researcher decided to visit a number of fraternity parties and survey the students at those parties, he might arrive at a very different estimate of students' drinking habits. You might think that you would never use such an obviously biased sample to estimate the drinking habits of a population of college students. But, as you will see in a moment, people seem to make inferences in their everyday lives on the basis of equally biased samples.

As implied by the example of the dissatisfied student, researchers are not the only ones who attempt to make inferences about population values on the basis of sample values. People do this all the time. This chapter will examine a variety of areas in which we attempt to use sample data to draw conclusions about populations in our everyday lives. For instance, we will examine how the process of estimation applies to the conclusions we reach about the (a) events or people we see in the mass media, (b) causes behind people's behavior, (c) degree to which other people share our habits and preferences, and (d) extent to which others notice our embarrassing or brilliant moments. In each area, we will examine some interesting research studies that suggest that we sometimes use biased samples to make estimates of population values. If we fail to adequately consider, or make adjustments for, the bias in our samples, the resulting estimates will be inaccurate.

Biased Samples from the Mass Media

Chapter 2 showed how people tend to overestimate the percentage of reported crimes that are violent crimes. We also saw that people tend to overestimate the frequency of sensationalized, dramatic causes of death (e.g., accidents, tornadoes, and homicides) and underestimate the frequency of causes of death that are less spectacular (e.g., diabetes, stomach cancer, and stroke). People's estimates of these risks might be distorted by what they hear from the news media. For example, we might underestimate the frequency with which people die from unspectacular, or nonviolent, causes because these causes tend to be underreported by the media (Combs & Slovic, 1979). This phenomenon is directly related to our discussion of using biased samples to make inferences about population parameters. News reporters do not select a random sample of deaths to report. Instead, they tend to report deaths that resulted from sensational causes and focus less on deaths that resulted from less interesting causes. As a result, we end up with a biased sample on which to base our population estimates of the frequency of various causes of death. If, when making our population estimates, we do not adequately consider or adjust for the bias in the sample we have obtained from the media, then we end up with biased estimates.

Our perceptions of ourselves and others are also susceptible to the potentially biasing influence of the mass media. For instance, there seems to be no shortage of physically attractive men and women in television shows, movies, and popular magazines. Advertisements for beauty products, for example, have been criticized for their heavy reliance on thin, beautiful models (Richins, 1991). If you watch television shows or read magazines that have beautiful actors or models, what impact does this have on your perceptions of yourself and of your romantic partner? A number of studies have found that men who are shown photographs of physically attractive women later report being less attracted to their female partners and rate their relationships less favorably (e.g., Kenrick, Gutierres, & Goldberg, 1989; Kenrick, Neuberg, Zierk, & Krones, 1994).

Researchers have also found that exposure to physically attractive photographs of other women can affect women's perceptions of themselves. For example, Richins (1991) found that women who were exposed to advertisements with attractive models, as compared with women who saw advertisements without the models, rated themselves less *satisfied* with their own attractiveness (but not less *attractive*). Gutierres, Kenrick, and Partch (1999) obtained similar results. They asked participants to read eight profiles of same-sex individuals that included a photograph of a person who had either a high or low level of attractiveness. Gutierres et al. found that women who were exposed to photos of highly attractive women rated themselves less desirable as marriage partners than did women who saw the photos of less attractive women.[1] Interestingly, similar to Richins, Gutierres et al. found that viewing photos of highly attractive women did not affect the women's ratings of their own attractiveness. In other words, they did not perceive themselves to be any less attractive after seeing the attractive photos, but they did believe they were less desirable as marriage partners. Gutierres et al. suspected that this happened because viewing the attractive photos may have affected women's perceptions of the physical attractiveness of the population of other women. Viewing the photos of highly attractive women might have distorted (i.e., increased) the participants' estimates of the number of very attractive women who are available to men seeking mates, thereby lowering their own perceived desirability as marriage partners. This fits well with our discussion of the effect of sample bias on population estimates. Being exposed to a biased sample of other women might have biased the participants' estimates of the percentage of attractive women in the population.

The women in the study conducted by Gutierres et al. (1999) may not have experienced any long-lasting or far-reaching effects as a result of their temporary exposure to a biased sample of beautiful people. But one has to wonder what effect everyday exposure to very thin women in fashion magazines and advertisements has on the women in our society. Some authors believe this exposure might contribute to the incidence of eating disorders among women (e.g., Stice & Shaw, 1994).

---

[1]Gutierres et al. (1999) also included men in their study. They found that the men's ratings of their desirability as a marriage partner were not affected by the attractiveness of the male photos they viewed. However, the men who read descriptions of socially dominant men, as opposed to nondominant men, rated themselves as less desirable as marriage partners. Gutierres et al. explained that women (more than men) tend to value social dominance in a mate, whereas men (more than women) tend to value physical attractiveness in a mate. Thus, people's ratings of their own desirability as a marriage partner are most affected by the degree to which they believe members of their own sex have the qualities that potential mates desire.

Attribution: Determining Cause

Attribution is another area in which we sometimes rely on biased samples to make estimates of population values. **Attribution** involves determining the causes behind behaviors or events. For example, if you invite a friend over for dinner and he or she arrives one hour late, you would probably try to determine what caused your friend to be late. Did your friend get caught in traffic or experience car problems (i.e., a situational, or external, cause)? Or is your friend a person who is chronically late (i.e., an internal, or dispositional, cause)?

Consider some attributions you might have made in your college classes. Think of a professor you recently had for a class. How would you characterize your professor's personality? Would you say this professor is a carefree, fun-loving individual? What do you think this professor does in his or her free time? What types of television shows do you think your professor watches? You might think you know this professor quite well, especially if you have had him or her for several courses. However, I would bet that you do not know him or her as well as you think you do. Ask your professor some of these questions and see if you get the same answers you gave. Why do you think I believe that many students do not know their professors as well as they think they do? Before I answer that question, let's take a look at a classic study on the topic of attribution.

Imagine being in a study in which you are paired with another student to participate in a quiz game. Your partner is assigned the role of questioner, and she is asked to generate 10 challenging, general knowledge questions (e.g., "What is the capital of South Dakota?" "What does the 'B. F.' in B. F. Skinner stand for?"). She asks you these questions, and you can only answer only four of them. Afterward, you are asked to rate yourself and your partner on general knowledge using a scale ranging from 1 ("much worse than average") to 100 ("much better than average"). What rating would you give yourself? What about your partner? Ross, Amabile, and Steinmetz (1977) had college students participate in this type of quiz game. Students participated in pairs, with each member of the pair randomly assigned to the role of questioner or contestant. The questioners asked the questions they had generated, and contestants tried to answer them. Afterward, everyone was asked to rate themselves and their partners on general knowledge. The results showed that the contestants correctly answered an average of only 4 of the 10 questions. Moreover, the contestants rated themselves lower on general knowledge than they rated the questioners. In other words, the contestants believed that the general knowledge of the questioners was superior to their own. The questioners, on the other hand, did not perceive their own level of general knowledge to be very different from that of the contestants.

Remember that the students in this study were randomly assigned the roles of questioner and contestant. Thus, on the average, there should be no difference in questioners' and contestants' general knowledge. In fact, the questioners and contestants in this study performed similarly on a general knowledge quiz Ross et al. gave the participants near the end of the study. Why, then, would the contestants conclude that the questioners' general knowledge was superior to their own? Apparently, in making their attributions, the contestants failed to make an adequate allowance for the large situational advantage held by the questioners. Because the questioners were allowed to generate their own questions, they could choose questions that capitalized on their unique knowledge. They could choose questions that very few others could answer correctly. Thus, contestants were presented with a very unrepresentative, biased sample of the questioner's knowledge. Contestants proceeded to use this biased sample to estimate the questioner's knowledge and did not make an adequate adjustment for the sample bias. They attributed the questioner's ability to generate difficult questions to his or her superior knowledge (i.e., a dispositional attribution) as opposed to the situational advantage held by the questioner.

As it turns out, people have a pervasive tendency to draw inferences about a person's dispositions from behavior that can be completely explained by situational factors. This phenomenon is called the **fundamental attribution error** or **correspondence bias** (e.g., Gilbert & Malone, 1995; Ross, 1977). A real-life example that closely parallels the study by Ross et al. (1977) is the tendency of some people to conclude that professional actors' personalities are very similar to the characters they portray in movies and television shows. We may conclude that actors who consistently play characters who are witty, intelligent, or aggressive are truly like those characters in their everyday lives. However, a good actor's behavior might be mainly a function of the script (i.e., a situational cause) he or she is following. As with the questioners in the study by Ross et al., actors may be showing us a very biased, unrepresentative sample of their behavior. If we use this sample to estimate the actors' off-screen personalities, we may end up with biased estimates.

To return to our discussion of students' views of professors, Ross et al. (1977) pointed out that, in academic settings, professors are often in the advantaged role of questioner and students in the disadvantaged role of answerer. Those who witness students struggling to answer difficult questions posed by professors in class may come away with a biased sense of the professors' knowledge. As I implied earlier, I suspect that students might also develop a biased, or at least narrow, view of their professors' personalities. I would guess that many of my students have a fairly inaccurate, or incomplete, sense of my personality. Some of my students have expressed surprise when I tell them that I played guitar in several rock bands when I was in high school and college. They also seem surprised when they learn that I like to cook a variety of ethnic foods (e.g., Thai, Ethiopian, Indian). The problem is that the impressions they have of me were formed in a rather restricted environment. They only see me in a limited variety of situations, mainly the classroom. A professional setting such as this requires a person to restrict, at least to some extent, his or her range of behaviors. When was the last time you saw a professor on campus jamming on a guitar, watching a soap opera, playing the part of a bag lady in a play, dancing to a rock song, or playing silly games with his or her kids? Some professors do these things, but you may never see that side of them. Instead, you are exposed to a very limited sample of their behavior. As a result, you may end up with a biased estimate of the population of behaviors that make up who they are.

The topic of attribution is also relevant to how individuals determine who is responsible for various outcomes or events. For example, think about a roommate, friend, or spouse with whom you currently live or used to live. Who do you believe had the most responsibility for each of the following activities (choose whether the responsibility was primarily yours, primarily your roommate's, or equally shared): cleaning dishes, cleaning house, shopping for groceries, planning joint leisure activities, causing arguments between the two of you, making the house messy, and taking out the garbage? Do you believe that you had most of the responsibility for these activities? Do you think your roommate would agree with your judgments on these items? Have the two of you ever gotten into arguments about who contributes more to household chores?

Ross and Sicoly (1979) suspected that people have an egocentric bias in the way that they attribute responsibility for joint ventures or products. In one study, they asked husbands and wives to estimate the extent of their responsibility for each of 20 activities (e.g., cleaning house, making the house messy, taking out the garbage, and causing arguments) . Most of the couples overestimated their personal contributions to these activities (i.e., the estimates of each couple typically added up to more than 100% of the response scale). Why would people overestimate their personal responsibility for such activities? One reason may be that people base their estimates of the entire population of contributions to an activity on a biased sample of those contributions. This sample may be biased in favor of their own contributions, because they may be more aware of and more likely to remember their own—as opposed to another person's—contributions. Consistent with this explanation, Ross and Sicoly found that the spouses in their study were able to recall more examples of their own—as compared with their partner's—contributions to the 20 activities they examined, and those who recalled more of their own contributions were more likely to overestimate their contributions.

Estimation Based on Single Cases

Previous examples have illustrated how individuals' estimates of population values might be biased by the use of unrepresentative samples that contain more than one observation or case (e.g., the exam grades of two students). However, as shown in Chapter 3, people sometimes also make ill-advised generalizations from single cases. For example, we might conclude that a weight-loss product is effective on the basis of a testimonial from a famous person who tried it and lost weight. Chapter 3 discussed the fact that such generalizations are problematic because they are based on an extremely small sample, and there may be alternative explanations for the weight loss. Another problem with making inferences based on single cases is that a single case may not be representative of the population from which it came. In other words, we do not know whether the effect of a weight-loss product on one person is typical of the effect it would have on most people. Perhaps the manufacturers of this product had hundreds of people try it and found that it worked on only one person. They then proceeded to make a commercial for their product featuring the testimonial of this one person. Obviously, it would be a huge mistake for us to conclude, after watching this commercial, that this product is effective for most people. Nevertheless, Hamill, Wilson, and Nisbett (1980) demonstrated that people can sometimes be surprisingly insensitive to the problem of sample bias when faced with single-case evidence.

In one study, Hamill et al. (1980) had participants read a description of one woman on welfare. She was described as a middle-aged, obese, irresponsible, Puerto Rican woman who had been on welfare for a long time. Moreover, the description explained that her home was full of filth and cockroaches. For some participants, this woman's long stay on welfare was characterized as atypical of welfare recipients, and they were told that middle-aged people stay on welfare for an average of 2 years. For a second group of participants, the woman's lengthy stay on welfare was characterized as typical; they were told that the average length of time on welfare was at least 8 years. A control group of participants did not read the description of the woman. Then, all of the participants were asked a number of questions to assess their attitudes toward welfare recipients in general. Interestingly, the participants who had read the negative description of one welfare recipient reported more unfavorable attitudes toward welfare recipients in general than did the participants who had not read the description, and this occurred regardless of whether they were told that the woman's stay on welfare was typical or atypical.

In a second study, Hamill et al. (1980) showed participants a videotape of a prison guard who discussed his job at a state prison. One group of participants saw a guard who appeared humane, compassionate, and interested in rehabilitating prisoners. A second group of participants saw a guard who appeared inhumane and unconcerned about the welfare of the prisoners. Moreover, some of the participants in each group were told that the guard's humaneness (or inhumaneness) was very atypical of the prison guards at the state prison, some were told that it was quite typical, and others were told nothing about the typicality of the guard. Afterward, participants were asked to report their views on American prison guards in general. Of course, participants who were told that the guard they saw was typical of the prison guards at a prison would have some justification for making inferences about all guards on the basis of this one guard, but the other participants would not. In fact, the participants who were told that the guard was atypical would have no justification whatsoever for inferring that all prison guards were similar to the guard they saw. Nevertheless, the participants in all three sampling conditions made similar inferences about prison guards in general. Compared with participants who saw no interview with a guard, the participants who saw the interview with the humane guard rated all guards more positively, and the participants who saw the inhumane guard rated all guards less positively. Thus, the participants' opinions about all prison guards were influenced by the interview with one prison guard, even when they were told that guard was a biased, atypical sample.

Chapter 3 discussed how people sometimes violate the law of large numbers by choosing single-case evidence over group statistics when drawing conclusions. Hamill et al.'s (1980) participants not only demonstrated a willingness to draw ill-advised inferences from single cases, but they also did so when they knew these cases were unrepresentative of the population from which they were drawn. The combination of the extremely small sample size and admittedly biased nature of that sample made these inferences especially unwarranted and shocking.

We may use biased, single-case samples to make inferences in our daily lives in many ways. For example, imagine visiting a new restaurant. Because you are unfamiliar with their food, you ask the waitress to recommend a dish. She suggests the best dish on the menu, and you absolutely love it. You then proceed to tell all of your friends about the wonderful food at this new restaurant. Unfortunately, your friends visit this restaurant and inform you that they were not at all impressed with the food. What went wrong? It seems that you may have fallen victim to the same mistake in statistical reasoning that shocked you when you read about the participants in the study by Hamill et al. (although the mistake in this example is not quite as blatant). The waitress recommended an item (i.e., a sample of $n = 1$) from the menu (i.e., the population) that she knew was an unrepresentative, or unusually delicious, sample. You proceeded to use that biased sample to estimate the quality of the entire menu. However, the menu items you did not sample were not as good as the dish you tried. Thus, when your friends went to the restaurant and tried these other menu items, they were not impressed with the quality of the food.

I must admit that the temptation to form unjustifiable, sweeping generalizations after trying a single dish at a restaurant can be overwhelming. I recently visited a Sri Lankan restaurant in Cincinnati and tried one of their main dishes. I thought it was mediocre, and I began to tell my friend that I was not very impressed with Sri Lankan food. Luckily, I stopped myself and acknowledged that my impression was based on a small sample. It would be unwise to use this sample to make an inference about the quality of food at this restaurant, let alone the quality of Sri Lankan food in general! Indeed, my friend replied that he had been to a Sri Lankan restaurant in Minnesota, and he thought the food was wonderful. However, because he knew that this restaurant had an award-winning chef, he admitted that his sample of Sri Lankan food may be biased. Perhaps both of us should wait until we have had a larger, more representative sample of Sri Lankan food before drawing any firm conclusions about the typical quality of Sri Lankan food in the United States.

Generalizing from the Self to Others

Hamill et al.'s (1980) demonstration of our willingness to form generalizations about a population of people based on our knowledge of only one individual, even when we have been told that individual is atypical, is indeed fascinating. Some equally fascinating phenomena arise when the individual that one uses as a basis for generalization is oneself. Let's take a look at three such phenomena: the false consensus effect, the illusion of transparency, and the spotlight effect.

The False Consensus Effect

Imagine that you have volunteered to participate in a study on your college campus on the topic of communication techniques. The experimenter explains that they are interested in studying the persuasiveness of various messages as a function of how they are presented to people (i.e., telephone calls, newspaper ads, handbills, etc.). You are told that today they are studying how people respond to messages that are presented on sandwich board signs. The experimenter explains that they would like you to wear a sandwich board sign that says "Eat at Joe's," walk around campus for 30 minutes, and keep track of how many people say something to you. Finally, the experimenter informs you that you do not have to participate in this study if you don't want to, but you will miss out on an interesting experience and the chance to do the researchers a big favor. Would you agree to wear this sign? What percentage of your peers do you think would agree to wear the sign? What percentage would refuse to do it?

Before continuing with this example, please answer a few more questions. Please indicate what percentage of college students you think engage in the following activities:

1. Smoke marijuana at least once a month _____%
2. Drink alcohol at least 3 times a week _____%
3. Study at least 3 nights a week _____%
4. Date on a regular basis _____%
5. Exercise at least 2 hours per week _____%
6. Snack between meals _____%

Now, go back through those six items and note whether you engage in each of the activities yourself.

The example involving the sandwich board sign describes an experiment conducted by Ross, Greene, and House (1977, Study 4). They asked some Stanford University undergraduates if they would be willing to wear a sign that read "Eat at Joe's"; other students were asked to wear a sign that read "Repent." Did you say you would wear the "Eat at Joe's" sign? Did you think that most of your peers would make the same decision? Ross et al. found that the students who *agreed* to wear this sign estimated that a majority of their peers (61%) would make the same decision; those who *refused* to wear the sign gave a much lower estimate (43%). Similarly, those who *refused* to wear the sign estimated that a majority of their peers (57%) would also refuse; those who *agreed* to wear the sign gave a much lower estimate (39%) of the percentage of refusals. These results illustrate a phenomenon called the **false consensus effect**: our tendency to believe that others share our opinions and habits. In other words, we tend to generalize from ourselves to others.

The false consensus effect appears in a wide variety of judgments. For example, Sherman, Presson, Chassin, Corty, and Olshavsky (1983) found that adolescent smokers, compared with nonsmokers, gave higher estimates of the percentage of their peers who also smoked. Kulig (2000) asked students to estimate consensus for 10 items, including the 6 items listed earlier. Take a look at the estimates you gave for these 6 items and compare them to the following average estimates given by participants in Kulig's study who *did not* engage in these activities themselves:

1. Smoke marijuana at least once a month: 58.6%
2. Drink alcohol at least 3 times a week: 64.8%
3. Study at least 3 nights a week: 43.2%
4. Date on a regular basis: 47.7%
5. Exercise at least 2 hours per week: 45.6%
6. Snack between meals: 64.0%

Which of these activities do you engage in yourself? For these activities, were your estimates higher than those of Kulig's (2000) participants who did not engage in these activities? Kulig found that students who engaged in these activities gave much higher estimates—11 to 22 percentage points higher—than did those who did not engage in them. Once again, we see that people's own habits influence their estimates of what others do.

To some extent, inferring that others engage in the same behaviors and hold the same opinions that we do is rational, because most people (by definition) are in the majority. Nevertheless, the previous discussion of the problems with making population estimates based on a single individual and of the importance of basing our estimates on representative samples suggests that using our own behaviors to estimate the behaviors of others could be problematic. As shown by the studies on false consensus, it could give us a false sense of the number of others who behave as we do. If we believe, incorrectly, that a majority of other people behave as we do, we could develop false notions that we are normal and those who do not behave like us are unusual or abnormal. As a result, we might be tempted to make biased attributions about the personalities of those whose behavior differs from our own. For example, Ross et al. (1977) found that students who refused to wear a sign that read "Eat at Joe's"—compared with those who agreed to wear the sign—were more likely to make strong inferences about the personalities of individuals who supposedly wore the sign (i.e., inferences about these individuals' shyness, submissiveness, etc.).

The Illusion of Transparency

Have you ever told a lie to someone and felt as though they knew you were lying? Have you ever tried to hide your disgust over a meal someone made for you and felt as though they could tell you disliked it? Are your internal reactions as transparent to others as you think they are? In the previous section, we considered how the false consensus effect results from our tendency to generalize from ourselves to others. We also generalize from ourselves to others when we use our awareness of our internal states to estimate the degree to which others notice these states. Researchers (e.g., Gilovich, Savitsky, & Medvec, 1998) have discovered that we have a tendency to overestimate the degree to which others notice our internal states. This phenomenon is called the **illusion of transparency.** Let's take a look some of the research on this interesting phenomenon.

In a study conducted by Gilovich et al. (1998), participants were directed to tell truths and lies—about famous people they had met, foreign countries they had visited, and so on—to a group of three to four other participants. Those who told the lies were asked to estimate how many of the other participants would be able to detect their lies, and these estimates were compared with the actual number of participants who detected the lies. They found that the liars estimated that 49% of the others detected their lies. However, only 26% of the observers actually identified the lies (an accuracy rate that was no better than random guessing). Thus, it seems that the participants felt more transparent than they actually were.[2]

In a second study, Gilovich et al. (1998) found that the illusion of transparency also occurs in situations in which we try to hide our disgust from others. They asked participants to maintain a neutral facial expression while tasting drinks, some of which contained a vinegar brine solution used for pickled grape leaves. They found that the participants overestimated the number of observers who would be able to detect their disgust. Thus, it appeared that the tasters felt more transparent than they actually were.

Gilovich et al. (1998) explained that the illusion of transparency results from an anchor-and-adjustment process in which individuals use their own awareness of their internal states as an anchor, or initial estimate of the degree to which others are aware of those states. Because they realize that other people do not share their perspective, they adjust this initial estimate downward to take this fact into account. However, as we saw in our discussion of anchoring and adjustment (p. 14), the adjustment is typically insufficient. Thus, the end result is an overestimation of the degree to which we are transparent to others.

[2] However, the participants may have overestimated the detectability of their lies because they simply believed that lies are readily detectable, not because they felt transparent while telling their lies. Gilovich et al. investigated this alternative explanation by having observers watch the participants tell lies and truths to the other participants. The observers were told ahead of time which statements would be truths and which would be lies, and they were asked to estimate the detectability of the lies. The researchers reasoned that the observers would not have any feelings of transparency because they were not the ones telling the lies, but they would share participants' beliefs about the general detectability of lies. Thus, if the observers also overestimated the detectability of the participants' lies, this would shed doubt on the illusion of transparency explanation. The results showed, once again, that the liars overestimated the actual detectability of their lies, and the liars' estimates were much higher than those of the observers. Therefore, the liars' overestimation apparently stemmed from an illusion of transparency rather than any general beliefs about the detectability of lies.

The illusion of transparency has interesting implications for a variety of real-life situations. For instance, if you attempt to have a serious discussion with your romantic partner in order to resolve a conflict, you might become very frustrated and angry if your partner behaves in a very lighthearted, joking manner during the discussion. Your anger and frustration might be partly a result of your belief that your partner knows that you are trying to have a serious conversation. However, your motives might not be as transparent to your partner as you suspect (cf. Vorauer & Claude, 1998).

Another implication of the illusion of transparency is related to a phenomenon called the **bystander effect,** which is the tendency for people to be less likely to help a person in need when there are other people (bystanders) nearby at the time of the emergency. For example, Latané and Rodin (1969) staged an emergency in front of male students who were in a room filling out a questionnaire as part of a research study. While the men were working on the questionnaire, they heard a female researcher fall off of a chair in a nearby office. They also heard the woman scream, complain that her ankle was hurt, and act as if something heavy was on top of her. Latané and Rodin found that most of the male students tried to help the woman when there were no other male students in the room with them. However, when there was one other student in the room at the time of the emergency, they were much less likely to help. One reason for this effect is that we tend to look to others to determine if an event is an emergency— but, in these types of situations, people seem to prefer to appear poised and under control. The illusion of transparency leads us to believe that our alarm is apparent to other bystanders, even though we actually appear just as calm and poised as everyone else. Because the other bystanders are exhibiting much less alarm than we believe we are exhibiting, we might conclude that they are truly less concerned about the situation. As a result, we are likely to interpret the situation as a nonemergency and fail to help (cf. Gilovich et al., 1998).

The Spotlight Effect

Several years ago, I decided to shave off my mustache. I had worn that mustache for more than 10 years and thought it would be nice to make a change. Of course, after I shaved it I was extremely self-conscious when I went out in public. I felt sort of naked, and I was sure that most of the people who knew me would immediately notice the dramatic change in my appearance. Of course, some people did notice the change immediately, but today I wonder how much I overestimated the number of people who would notice that change in my appearance. Now that I think about it, some people I knew never said anything to me about my naked face. Did they ever notice? Should I have been as concerned as I was about what people would think about the change in my appearance? Did the social spotlight shine on me as brightly as I thought it did?

You have probably had the feeling that other people's eyes are on you, noticing your new haircut, the pimple on your face, the brilliant comment you made in class, the embarrassing food stain on your pants, or the stylish new clothes you just bought. Do other people really notice these things as much as we think they do? This question was the focus of a series of experiments conducted by Gilovich, Medvec, and Savitsky (2000). In one study, they required some Cornell University students to put on an embarrassing T-shirt that displayed a large picture of Barry Manilow. Then, they asked each student to enter a room that contained two to six other students sitting at a table facing the doorway. Afterward, the students who wore the T-shirts were asked to estimate how many of the observers in the room would be able to recall who was on their T-shirt. They found that the students dramatically overestimated the number of observers who would be able to recall who was on the T-shirt. On the average, the students estimated that 46% of the observers would remember the T-shirt, but only about 23% actually did. Gilovich et al. (2000) labeled this phenomenon the **spotlight effect**—our tendency to overestimate the extent to which other people notice our behaviors and appearance.

Gilovich et al. (2000) found that the spotlight effect occurs in a variety of contexts, both embarrassing and nonembarrassing. In a second study, they had participants wear nonembarrassing T-shirts with pictures of famous people, and they found that people still overestimated the extent to which others noticed their shirts. In a third study, they asked students to engage in a discussion with a small group of students. Then, they asked the students to estimate how highly they would be ranked by the other members of their discussion group on their positive (e.g., advancing the discussion) and negative (e.g., speech errors) contributions to the discussion. As before, they found that the students' estimates were higher than the actual judgments made by the other group members.

The spotlight effect and the illusion of transparency are somewhat different phenomena; the former concerns the extent to which we believe others are aware of our outward actions and appearance, and the latter concerns the extent to which we believe others are aware of our internal states. However, the two phenomena may be produced by the same underlying mechanism: anchoring and adjustment. In both cases, we seem to anchor our estimates on our own vivid experiences and then adjust, insufficiently, to account for the fact that other people do not have the same perspective we do. When estimating the extent to which other people will notice an embarrassing shirt, for example, we might start with how noticeable we feel the shirt is and then adjust from there.

The anchor-and-adjustment explanation for the spotlight effect received some support from another study conducted by Gilovich et al. (2000). They reasoned that, with the passage of time, a person should adapt to, and become less concerned about, the embarrassing shirt he or she is wearing. As result, the person would have a lower initial estimate (i.e., anchor) of how noticeable the shirt is to others, and this would result in a final estimate that is closer to reality. To test this idea, they had some participants walk into a room of observers 15 minutes after they had put on an embarrassing T-shirt; other participants walked into the room immediately after putting on their T-shirts. As they predicted, those who waited (compared with those who did not wait) before entering the room gave much lower estimates of the percentage of people who would be able to recall who was on their shirt. Finally, Gilovich et al. suggested that, if we have adapted to some aspect of ourselves to the degree that we no longer notice it as much as others do, a reverse spotlight effect might occur. For example, if we have adapted to the smell of our body odor or cologne, we might underestimate the extent to which other people notice the scent.

Both the illusion of transparency and the spotlight effect provide additional examples of the problem of using a biased sample of one individual to make estimates of population values. However, in these cases, the individual who is sampled (oneself) is not even a member of the population we are estimating! As was demonstrated by the research on the illusion of transparency and the spotlight effect, our perspective may be quite different from that of others. Thus, if we want to get a good sense of how other individuals perceive us, we should draw our sample from them rather than ourselves.

Summary

The process of estimation—that is, using sample values to estimate population values—is relevant to our everyday lives. We have seen how people might use this process to estimate a variety of population values, including the frequencies of various causes of death, the physical attractiveness of people, the personalities of professors, the quality of food at a restaurant, the degree to which others are like us, and the extent to which our emotions and actions are noticeable to others. We have also seen how the estimation process used by people sometimes violates a very important, fundamental rule of statistical inference: If you want to make an accurate estimate of a population value, you should rely on a sample that is representative of the population from which it came. Our population estimates will be inaccurate if we base them on unrepresentative, or biased, samples and fail to adequately consider or adjust for this bias. Our reliance on biased samples in our everyday lives could lead us to make a number of inaccurate conclusions, such as those about the difficulty of an exam, the personality of another person (e.g., the correspondence bias), or the general level of attractiveness of women in our society. When we rely on ourselves as our sample, a number of interesting phenomena arise: We might conclude, incorrectly, that other people behave as we do (i.e., the false consensus effect), read our emotions (i.e., the illusion of transparency), or notice our brilliant or embarrassing moments (i.e., the spotlight effect).

● **PRACTICE PROBLEMS**

1. Researchers (e.g., Heath & Petraitis, 1987) have found that people who watch crime dramas on television are more likely than those who do not watch such dramas to fear cities such as New York City, but they are not more likely to

fear their own neighborhoods. Explain how both results are related to the process of estimation and to the problem of sample bias.

2. Students sometimes complain that a professor talks "over their heads." Explain how this example might be related to the process of estimation and to the problem of sample bias. (Hint: Assume that the students' perceptions are accurate and the problem lies with the professor.)

3. An accountant for a college is asked to give a presentation on the annual budget to the college's administrative officers. Because she is not used to giving presentations, she is very nervous. Throughout the presentation, she keeps thinking, "This is so embarrassing; these people can see that I am a nervous wreck." After the presentation, the administrative officers discuss how impressed they were that the accountant appeared so calm and collected during her presentation. Why did the accountant have such a distorted view of the officers' perceptions of her? What specific concept is illustrated by this example?

4. Linda is an avid fan of her college's football team. She has watched every game they played over the past 3 years. One night, she attends a party and sees some of the football players there. A friend of hers informs her that one of the linebackers at the party noticed her, and he is thinking of asking her out on a date. Linda has seen this particular linebacker play a number of games, and she remembers he has a hard-hitting style of playing and likes to "talk trash" on the field. When her friend asks Linda what she will do when he asks her out on a date, Linda says, "I like calm, sensitive guys; I think he would be a bit too rowdy and aggressive for my taste." But when he actually asks her out, she accepts his invitation. After several dates, she admits to her friend that she is surprised that he is so calm and sensitive. She has also met several of his friends, and they have confirmed her new impression of him. What might account for her faulty initial impression of the linebacker? What specific concept is illustrated by this example?

Correlation

> Information about the relationships or covariations between events in the world provides people and animals with a means of explaining the past, controlling the present, and predicting the future, thereby maximizing the likelihood that they can obtain desired outcomes and avoid aversive ones.
>
> *Alloy and Tabachnik (1984)*

Several weeks after my son, Ryan, was born, my wife wondered why he sometimes cried out loudly for no apparent reason; he seemed to be in pain. One time after he cried out, she noticed that his diaper was wet. She suspected that he was experiencing painful urination and proceeded to test her hypothesis by checking his diaper every time he cried out in this manner. Sure enough, his diaper was wet on many of the occasions she checked him. What would you conclude at this point? Assume that she came to you for advice and told you that his diaper was wet on eight of these occasions. Would you advise her to call Ryan's doctor? Do you need any additional information before making this decision? If so, what information do you need?

My wife did call the doctor and took Ryan in to be tested. The doctor also suspected something might be wrong, and he collected a urine sample (a procedure that was a bit painful itself) with which to conduct some tests. The results were negative; nothing was wrong with him. Why, then, did it seem that there was a relationship (i.e., a correlation) between his crying and diaper wetting?

Assume that over the period of several weeks, my wife had checked Ryan's diaper every time he cried loudly. Let's say he cried like this 10 times during that period, and on 8 of those occasions his diaper was wet. Does that sound like convincing evidence of a relationship between his crying and diaper wetting? Would you, like my wife, be concerned that something was wrong with your baby after making such observations? I would bet that most parents would be concerned at that point. However, it is entirely possible that these observations reflect no relationship whatsoever between his crying and diaper wetting. How could that be, you ask?

To determine if there is a relationship between these two events, we need more data. We have seen that his diaper was wet on 8 out of 10 (80%) of the occasions that he cried. But, in order to accurately determine if there is a correlation between

Table 5.1

Hypothetical Frequencies of Crying and Diaper Wetting

	Diaper Wet?		
Crying?	Yes	No	Total
Yes	8	2	10
No	8	2	10
Total	16	4	20

his crying and diaper wetting, we also need to know how often his diaper was wet when he was *not* crying. Imagine that we checked Ryan's diaper on 10 randomly chosen occasions during which he was not crying, and we wrote down whether his diaper was wet. Table 5.1 shows the complete set of hypothetical observations in the form of a 2 × 2 contingency table, or four-cell matrix. The top row indicates what happened when we checked his diaper after he cried, and we have already discussed the results of those observations. The second row indicates what happened when we checked his diaper when he was not crying. As you can see, his diaper was wet on 8 out of 10 (80%) of the occasions during which he was not crying. Thus, it seems that no matter when his diaper was checked, it was wet 80% of the time. Apparently, he wet his diaper frequently, but these data clearly indicate that his diaper wetting was not related to his crying. These two events are uncorrelated, or independent of one another.

Of course, the data in Table 5.1 are hypothetical. But I wonder if my wife would have observed a similar pattern of data if she had collected a more complete set of observations to determine the correlation between crying and diaper wetting. In her defense, however, I doubt that many people would conduct such a careful test of their everyday hypotheses about relationships between events or variables. As we will see later, many people fail to realize that all four cells in a contingency table (like the one in Table 5.1) are needed to determine the correlation between two events. After reading the first paragraph of this chapter, did you say that you needed additional information before giving your advice? If so, did that additional information involve checking his diaper when he was not crying? Did that information even seem relevant to you?

This chapter will examine the topic of **correlation,** first by discussing what a correlation is and examining how the concept of correlation is important to scientific researchers. Then, we will discuss how the concept of a correlation is relevant to our everyday lives. Our discussion will then focus on a number of ways we might develop inaccurate estimates of correlations, including the use of incomplete data, testing our hypotheses with a positive-test strategy, allowing our biased expectations to unduly influence our judgments, and overreliance on distinctive or rare cases. We will also examine how stereotypes and illusions of control are related to our ability to assess correlations. Finally, we will discuss whether people can ever accurately assess correlational relationships.

What Is a Correlation?

When two events or variables are correlated, they are associated with one another. For example, an increase in the temperature outside is associated with an increase in the occurrence of violent crimes (e.g., Anderson & Anderson, 1984); therefore, temperature and violent crime are correlated. More specifically, the relationship between temperature and violent crime is a **positive correlation,** because an increase in temperature is associated with an increase in violent crime (i.e., changes in the two variables occur in the same direction). If an increase in temperature were associated with a decrease in violent crime, this would be a **negative correlation** (i.e., changes in the two variables occur in opposite directions). Researchers often calculate a **correlation coefficient** to measure the strength of a correlation. Correlation coefficients—often symbolized by the letter r—range in value from -1 to $+1$, and coefficients closer to either value of 1 (i.e., either -1 or $+1$) indicate stronger correlations. Thus, a correlation coefficient of $r = .9$ (or $r = -.9$) is quite strong, $r = .2$ is weak, and $r = 0$ indicates no correlation at all.

Researchers examine correlations between events and variables in order to enhance their ability to understand, predict, and control events. For instance, if research studies indicate that high school students' scores on the Scholastic Aptitude Test (SAT) are positively correlated with their performance in college, we can predict that students who perform well on this test will also tend to perform well in college. Moreover, colleges might gain some control over the quality of incoming students by accepting only students who scored at a certain level on the SAT. As another example, researchers interested in human aggression might enhance their understanding of aggressive behavior by determining what variables are correlated with this behavior. If we discover that aggressive behavior is more likely when people are frustrated, experiencing pain, or in a hot environment, then we gain a better understanding of the psychological and environmental conditions under which aggressive behavior occurs.

Assessing Correlations in Everyday Life

As suggested by the opening quote for this chapter, our ability to detect relationships or correlations between events in our everyday lives is an important one. As is the case with scientific researchers, our ability to accurately assess correlations between everyday events and variables may enhance our ability to understand, predict, and control events in our everyday lives. There is a wide variety of relationships between events or variables that might be important for us to detect. For example, it might be important to know whether (a) your town has a higher crime rate at night than during the day, (b) you will get higher grades on exams if you put in more study time, (c) your nose feels more congested when the pollen count is high outside, (d) you are more likely to roll a seven at the craps table when you blow on the dice before tossing them, and (e) your child's behavior improves after you discipline him or her. These are just a few examples of the many ways that the concept of correlation is relevant to our everyday lives.

A common type of correlation that we might assess in our everyday lives is whether or an event or attribute occurs more often in the presence than in the absence of a second event or attribute (cf. Gilovich, 1991). For example, recall Question 4 from the questionnaire in Chapter 1:

4. A man suspects that blond women tend to be especially talkative compared with other women. Assume that, over the period of a year, he met and talked to 40 blond women who were very talkative. Thus, he concludes that there is indeed a relationship—or correlation— between the attributes "blond" and "talkative." Do you think he is correct?

How did you answer this question?

If a psychologist wanted to investigate this relationship, she would be careful to record observations from all four cells in the 2 × 2 contingency table that is often used to represent relationships between dichotomous variables (i.e., variables that have only two values). In other words, she would record observations of women who are blond as well as those who are not blond and of women who are talkative as well as those who are not talkative. Imagine that a psychologist made these observations. Hypothetical data from observations of 87 women are shown in Table 5.2; note that I labeled the four cells A, B, C, and D. As before, 40 blond women were talkative. Now

Table 5.2

Hypothetical Frequencies of Blondness and Talkativeness Among Women

Blond?	Talkative?		Total
	Yes	No	
Yes	40 (A)	20 (B)	60
No	18 (C)	9 (D)	27
Total	58	29	87

that you have the complete set of observations, can you conclude that there is a correlation between blondness and talkativeness? As implied by the wet diaper example, one way to assess this correlation is to compare the proportion of all blonds who are talkative [A/(A + B) = 40/(40 + 20) = 2/3, or .67] with the proportion of all nonblonds who are talkative [C/(C + D) = 18/(18 + 9) = 2/3, or .67].[1] If these two proportions are similar, there is no correlation ($r = 0$) between the two variables. Thus, there is no correlation between blondness and talkativeness in these hypothetical data.[2]

Difficulties in Assessing Correlations

Assessing Correlations from Incomplete Data

A common mistake people make in assessing correlations between dichotomous variables is failing to realize that data from all four cells in a contingency table are necessary for the task (e.g., Alloy & Tabachnik, 1984; Nisbett & Ross, 1980). One common strategy people use is to focus mainly on cell A, the yes-yes cell (Kunda, 1999; Smedslund, 1963; Ward & Jenkins, 1965). The example of the man who suspected that blond women are very talkative illustrates this strategy. After observing 40 blond women who were talkative, he concluded there was a correlation between blondness and talkativeness. Of course, the data from all four cells in Table 5.2 indicate there is no correlation between these two attributes. Even considering cells A and B in Table 5.2—another strategy people sometimes use—can be misleading. When one hears that 67% (or 2/3) of blonds were talkative, it may be tempting to conclude that there is a relationship between blondness and talkativeness. However, the data in cells C and D reveal a similar percentage among nonblonds. Similarly, when we discussed the wet diaper example, we saw that Ryan's diaper was wet on 80% of the occasions on which cried, but it was also wet on 80% of the occasions on which he did not cry.

---

[1]These hypothetical data indicate that 2/3 of women are talkative, but, of course, this figure is fictional. I hope you did not conclude that these data imply that women are more talkative than men. If you did make such a conclusion, you would be exhibiting the type of mistake discussed in the next section of this chapter: using only cells A and B to assess a correlation. Because no data for men were presented, you do not have enough information to assess the (hypothetical) correlation between gender and talkativeness. To assess such a correlation, you would need to know how many observations were of talkative women (cell A), nontalkative women (cell B), talkative men (cell C), and nontalkative men (cell D). You only have information from cells A and B. The data from those two cells suggest that 2/3 of women are talkative, but they say nothing about men.

[2]Note that a psychologist would typically do more than a simple comparison of proportions to assess the correlation between these two variables. She might use a chi-square test for independence to determine whether a statistically significant correlation exists between these two variables. Alternatively, she might measure the strength of the correlation between the two variables by calculating the Pearson correlation coefficient—or, more specifically, the phi coefficient—for these data. Although a detailed discussion of these statistical tests is beyond the scope of this book, interested readers may want to verify the fact that neither statistical test would reveal a correlation between blondness and talkativeness for the data in Table 5.2.

Nisbett and Ross (1980) pointed out that people may use the strategy of focusing only on cell A when they assess the correlation between praying to God and experiencing a favorable outcome. When asked if God answers their prayers, people might say "yes" because they remember many times that they prayed for something and received it; these observations would fall into the yes-yes cell in a 2 × 2 contingency table. Similarly, Kunda (1999) indicated that people might support a claim that astrology is valid by giving examples of events predicted by their horoscope that actually came true; once again, these observations would fall into the yes-yes cell. Even if we know that prayers to God were answered more often than they were not answered, we cannot conclude God answers our prayers. These observations would fall into cells A and B in a contingency table, and we would still lack all of the observations necessary for determining the correlation. If we had observations from cells C and D, they might indicate that our wishes come true more often than not, even when we do not pray.

Nisbett and Ross (1980) pointed out that it might seem counterintuitive that the observations in cell D are necessary for assessing correlations. If you wanted to determine if God answers your prayers, would you realize that favorable outcomes that did not occur when you did not pray for them are relevant to this judgment? If you suspected that blond women are talkative, would you consider women who are not blond and not talkative before reaching a conclusion about this relationship? If you suspected that your baby's crying was a result of painful urination, would you count the number of times his diaper was dry when he was not crying? In each case, these observations are necessary for accurately assessing the suspected correlations. However, because such observations may seem irrelevant to us, we might overlook them.

The Positive-Test Strategy

Focusing on the yes-yes cell in contingency tables might be partly a result of the **positive-test strategy** we often use to test a variety of hypotheses (Fiedler, Walther, & Nickel, 1999; Klayman & Ha, 1987). The positive-test strategy involves searching for evidence that matches the hypothesis being tested, as opposed to evidence that would match an alternative hypothesis. For example, if I asked you to determine if another person is assertive, you might try to recall examples of assertive behaviors he or she exhibited in the past. If I asked you to determine if the person is unassertive, or deferent, you might try to recall examples of deferent behaviors he or she exhibited in the past (cf. Snyder & Cantor, 1979). Of course, most people have probably engaged in both assertive and deferent behaviors at some point during their lives. Thus, no matter which hypothesis you tested, you would probably be able to remember examples of the behaviors for which you were searching. As a result, regardless of the hypothesis you tested, you might conclude that your hypothesis was confirmed. However, if you conducted a more balanced search for information and tried to recall assertive as well as deferent behaviors, you would probably form a more accurate impression of the person.[3]

In a classic study concerning how people test hypotheses about others, Snyder and Swann (1978) asked their participants to test a hypothesis about the personality of a person whom they had never met. Some participants were asked to assess how extraverted, or outgoing, the target person was, and other participants were asked to determine how introverted, or shy, the target person was. The participants were allowed to choose 12 questions—out of a total of 26 questions that included questions about extraverted and introverted behaviors—to ask the target person in order to test their hypotheses. What types of questions do you think were selected by the participants who were asked to determine if the target person was an extravert? What questions were selected by the participants who were asked to determine if the target person was an introvert? As you might have guessed, Snyder and Swann found that participants were more likely to ask questions related to extraverted behaviors (e.g., "In what situations are you most talkative?" p. 1204) when they were testing an extraverted hypothesis as compared with an introverted hypothesis. Similarly, their participants were more likely to ask questions related to introverted behaviors (e.g., "What factors make it hard for you to really open up to people?" p. 1204) when they were testing an introverted hypothesis as compared with an extraverted hypothesis.

The participants in Snyder and Swann's (1978) first study did not actually ask another person the questions they selected. What do you think would happen if the participants were allowed to ask another person these questions? Would the participants' questions actually elicit responses that confirmed their hypothesis, regardless of whether they had an introverted or extraverted hypothesis? If someone asked you a series of mostly extraverted questions, would you appear extraverted to those who heard your answers? If you were asked introverted questions, would you appear introverted to those who heard your answers?

In a second study, Snyder and Swann (1978) allowed the participants to choose questions as before, but this time they were allowed to pose those questions to another person (a "target" person) whom they had never met prior to this study. The target persons were asked to simply answer the questions in an open, candid manner. To determine if the target persons' responses to these questions provided confirmation of the participants' hypotheses, Snyder and Swann had a separate group of people listen to an audio recording of the target persons' responses to some of the questions. They discovered that the target persons in the extravert-hypothesis condition were judged by the listeners to be more extraverted, confident, poised, and energetic than were the target persons in the introvert-hypothesis condition. Thus, the target persons' behavior in the interviews actually confirmed the participants' hypotheses.

[3]Note, however, that the positive-test strategy would lead to the same conclusion as a more balanced search in cases where the person's behavior is very uniform (e.g., the person is always assertive).

It seems likely that most of us have engaged in extraverted as well as introverted behaviors at some point. Therefore, when we are asked about our extraverted behaviors, we can generate examples; similarly, when we are asked about our introverted behaviors, we can generate examples. If another person asks us mainly about our extraverted behaviors, they might infer from our answers that we are indeed extraverted. Likewise, if a person asks us mainly about our introverted behaviors, they might infer from our answers that we are introverted.

Although the studies conducted by Snyder and Swann (1978) were not designed to investigate the role of the positive-test strategy in the detection of correlations, their results are certainly relevant to this topic. As mentioned earlier, the use of the positive-test strategy to test hypotheses about correlations could lead to a focus on the yes-yes cell in a contingency table, because observations in that cell are clearly consistent with the hypothesized correlation between two events. For example, if we suspect that blondness is correlated with talkativeness among women, we may be most likely to search for talkative blonds. Certainly, we would expect to find such women if this correlation exists. Talkative brunettes or redheads may not seem as relevant to our task of assessing this correlation but, as we saw earlier, they are indeed relevant. Similarly, if we suspect that a baby's crying is correlated with a wet diaper, we may be most likely to check for a wet diaper when he is crying; after all, we would expect his diaper to be wet on such occasions if these two events are related. Discovering that his diaper is dry when he is not crying might seem unimportant to our assessment of this correlation, but it is actually very important.

One way to avoid the biased conclusions that could result from the positive-test strategy is to consider a hypothesis that is the opposite of your original hypothesis (cf. Lord, Lepper, & Preston, 1984). For example, if you suspect that a person is extraverted, ask yourself if there is also evidence that the person is introverted. If you suspect that a baby's wet diaper coincides with his crying, ask yourself if it also coincides with his quiet moments.

Biased Expectations and Illusory Correlations

Clinical psychologists sometimes rely on psychological tests to diagnose their clients' problems, and they may rely on their clinical experience to interpret the results of these tests. During the late 1960s, Loren and Jean Chapman (1967, 1969) noticed a disturbing problem with clinicians' observations of the results of a couple of psychodiagnostic tests: the Draw-a-Person Test (DAP) and the Rorschach Inkblot Test. In the DAP, a client draws a picture of a person, and the clinician examines this drawing for features that might indicate psychological problems. There was fairly consistent agreement among the clinicians who used the DAP that people with certain psychological symptoms drew pictures with particular characteristics. For example, the clinicians believed that people who were paranoid or suspicious of other people drew pictures with atypical eyes and that men who were worried about their manliness drew broad-shouldered, muscular figures. The problem was that research studies that carefully examined the actual correlations between drawing characteristics and the psychological symptoms of the test-takers found no such correlations. In other words, even though experienced clinicians widely believed in the existence of these correlations, their beliefs appeared to be erroneous. The clinicians had developed **illusory correlations,** which are relationships that do not exist or relationships that are not as strong or in the same direction as one believes. How could these professionals, who had plenty of opportunities to observe the actual relationship between the characteristics of people's drawings and their symptoms, be so wrong?

A similar phenomenon was occurring with the Rorschach Inkblot Test, a test that requires clients to interpret a series of inkblots. Many clinicians believed that homosexual males were more likely than heterosexual males to report seeing anal content, feminine clothing, or sexual organs in the inkblots. However, scientific research that examined the actual correlations between the sexual orientation of Rorschach test-takers and their responses to the inkblots found no such correlations. This research did, however, find that homosexuals were more likely than heterosexuals to report seeing animal-human figures in one inkblot and monstrous or threatening creatures in another inkblot. Nevertheless, a survey of clinicians showed that they seemed to be largely unaware of these valid signs of homosexuality (Chapman & Chapman, 1969). Once again, one has to wonder how these professionals could be so wrong.

Chapman and Chapman (1969) suspected that the clinicians saw these illusory correlations because they expected to see them. In fact, because most of these illusory correlations represented seemingly logical connections between test responses (e.g., reporting anal content) and the attributes of test-takers (e.g., homosexuality), Chapman and Chapman thought that naïve observers might develop the same illusory correlations after observing a series of hypothetical test responses that reflected no correlations at all. To test this, they showed college students a series of Rorschach cards that contained the supposed responses and attributes of clients who took the test. Some of the responses reflected valid signs of homosexuality; others reflected invalid signs. Moreover, some of the cards included client attributes that implied homosexuality. These data were fabricated in such a way that there was no correlation between the attributes and responses of the clients. Nevertheless, the students reported seeing the same illusory correlations that had been reported by clinicians; many of them reported that responses involving anal content, feminine clothing, and genitalia were correlated with the attribute of homosexuality. In another experiment, Chapman and Chapman showed students similar materials. However, this time the data on the Rorschach cards reflected a strong correlation between homosexuality and the valid signs of it (i.e., responses indicating animal-human figures) and no correlation between homosexuality and invalid signs (e.g., responses concerning anal content). Nevertheless, the invalid signs were most frequently reported by students to be correlated with homosexuality.

Our use of the positive-test strategy may play a role in the development of illusory correlations. For example, if a person believes that homosexuality is associated with anal content in people's responses to Rorschach inkblots, he or she may be most likely to take note of those observations that match the suspected correlation and pay less attention to observations that are inconsistent with the correlation. This is similar to how the positive-test strategy might have led to illusory correlations in the wet-diaper and talkative-blonds examples.

Newer research has confirmed that our expectations have an important influence on the correlations we detect (e.g., Alloy & Tabachnik, 1984; Hoover & Milich, 1994). The research by Chapman and Chapman (1967, 1969) showed how people's expectations might bias their assessments of correlations from data that are comprised of unambiguous, pre-classified responses and attributes. However, a person's expectations might also affect their interpretations of more ambiguous responses from others. For instance, if you expect that the consumption of sugar makes a child hyperactive, you may be more likely than a person who does not hold this belief to characterize that child's behavior as hyperactive after he or she eats sweets. Hoover and Milich (1994) tested this possibility with mothers and their supposedly sugar-sensitive children. They noted that (a) a common belief among parents and physicians is that the consumption of sugar is associated with problematic behaviors among children and (b) scientific studies of the effects of sugar on children's behavior have failed to find any negative behavioral effects of sugar consumption. Thus, it appears that parents' and physicians' belief in such effects reflect an illusory correlation between sugar consumption and problematic behaviors.

Hoover and Milich (1994) wanted to see if a mother's beliefs about the sugar sensitivity of her child would influence her assessments of the child's behavior and her interactions with the child. In their study, they included only mothers who believed that their sons' behavior was affected by sugar. Some of the mothers were told that their son would be given a drink that contained a large dose of sugar, and the other mothers were told that the drink did not contain sugar (it contained aspartame instead). However, all of the drinks actually contained aspartame. Thirty minutes after the children consumed the drinks, the mothers were asked to play and perform some tasks with their sons while the researchers videotaped their interactions. They found that the mothers who believed their sons had a sugary drink rated their son's behavior as more hyperactive than did the mothers who believed their sons did not have sugar. However, the boys in the "sugar" condition were actually less active during the interactions than were the boys in the nonsugar condition (probably because the mothers in the sugar condition maintained closer contact with their sons during the interaction). Thus, parents' belief in an illusory correlation between sugar consumption and activity level may be a result of their biased perceptions of their child's behavior. If they expect that sugar consumption will increase their child's activity level, they may perceive their child's activity level to be higher after the child has consumed sugar than after the child has consumed foods without sugar. In reality, however, the child's activity level is not likely to have changed much as a result of sugar consumption.

Distinctive Events, Illusory Correlations, and Stereotypes

As we have seen up to this point, illusory correlations can occur in a variety of everyday situations, including psychologists' interpretations of people's responses to psychological tests and our observations of our children's behavior. One type of illusory correlation that may have particularly negative consequences is a **stereotype**—a belief about the traits or behaviors of a particular group of people. In other words, a stereotype is a perceived correlation between people's group membership and their traits or behaviors (cf. Fiedler et al., 1999). The belief that overweight people are lazy, jocks are stupid, or computer programmers are nerds are all examples of negative stereotypes. Of course, stereotypes can also be positive. For instance, people tend to believe that physically attractive people are also intellectually competent (e.g., Jackson, Hunter, & Hodge, 1995).

Because many stereotypes are inaccurate or overgeneralized, they are illusory correlations. These stereotypes might be formed in a manner similar to the other illusory correlations we have examined. For instance, if we have an expectation that overweight people are lazy, we might test this hypothesis by focusing mainly on observations of overweight people who match our hypothesis. Furthermore, our expectation might make us more likely to characterize an overweight person's behavior as lazy, whereas others might see the same behavior as normal. However, Hamilton and Gifford (1976) suggested a different mechanism—the co-occurrence of rare or distinctive attributes or events—that may also produce stereotypes. For example, seeing a gay male dressed in feminine clothes or a mentally disordered person murder someone might capture our attention and unduly influence our impressions of these groups, because these events involve the co-occurrence of unusual and distinctive events. People who have a mental disorder are fairly rare and distinctive, and the behavior of murdering someone is also rare and distinctive. When these two events co-occur (i.e., a mentally disordered person murders someone), we might be especially likely to notice and remember them, and they might form the basis of our generalizations about the behaviors of mentally disordered people in general. When people without mental disorders commit murders, these events might influence our judgment to a lesser degree because they are not quite as rare or distinctive. Of course, any time we hear about members of a minority group engaging in unusual behaviors, it might influence our impressions unduly, so this phenomenon is relevant to our beliefs about a wide variety of groups.

Hamilton and Gifford (1976) tested their hypothesis by showing a series of slides to the participants in their study. Each slide contained a description of a desirable or undesirable behavior performed by a hypothetical person. Moreover, each person was described as a member of either "Group A" or "Group B." For example, one description read, "John, a member of Group A, visited a sick friend in the hospital" (p. 394). Participants saw a total of 39 descriptions. The descriptions were designed so that

Group A was twice as big as Group B, and desirable behaviors were twice as numerous as undesirable behaviors. Thus, the descriptions of Group B members who performed undesirable behaviors were quite rare. However, both groups had the same ratio of desirable to undesirable behaviors; in other words, there was no correlation between group membership and the desirability of the behaviors. Nevertheless, Hamilton and Gifford predicted that the participants would form an illusory correlation and view Group B more negatively, because Group B was the smaller group and their negative behaviors were quite rare. Indeed, they found that the participants overestimated the number of negative behaviors performed by Group B members, and their overall impressions of Group B members were more negative than their impressions of Group A members. Thus, they had developed an illusory correlation between group membership and desirability of behavior that resulted in a negative stereotype of the members of Group B. Note that this stereotype was developed without any pre-existing expectations of these groups. Newer research has provided additional evidence for the biasing effects of distinctive or rare events (e.g., Hamilton & Sherman, 1994), but other researchers have argued that the effects observed by Hamilton and Gifford (1976) can also be produced by mechanisms other than distinctiveness (e.g., Chun & Lee, 1999; Fiedler, 1991).

Illusions of Control

Illusory correlations can sometimes produce another phenomenon called an **illusion of control,** In which we believe that we have more control over some event than we actually do. The perception of a connection, or correlation, between one's actions and the occurrence of an event is an important component of an illusion of control (Thompson, 1999). For example, a gambler might believe that blowing on the dice before tossing them onto the craps table will increase his or her chances of a desirable result. An athlete might believe that wearing her lucky baseball cap will improve her performance on the field. In each case, the illusion of a correlation between one's actions and the desired outcome may feed the illusion that one can control the outcome.

An incorrect belief that sugar consumption is correlated with problematic behaviors in their child (e.g., Hoover & Milich, 1994) might prompt parents to restrict the child's sugar consumption in an attempt to gain some control over the child's behavior. Although it may seem to the parents that the child's behavior improves after a restriction in sugar consumption, this relationship is illusory, and thus, so is their belief that they can control the child's behavior by altering what he or she eats.

In a well-known study, Langer (1975) demonstrated that people appear to believe that their actions can influence gambling outcomes that are completely determined by chance. She hypothesized that, if people are allowed to choose their own lottery tickets, they may perceive that their chances of winning are greater than if they had no such choice. The participants in her study were office workers who were either allowed to choose their own lottery ticket or were simply handed a ticket by a fellow worker who conducted the lottery. The participants were approached later and were told that a fellow employee wanted to buy a lottery ticket but no more tickets were available. Then, the participants were asked how much money they would sell their ticket for. Langer found that those who were allowed to choose their own ticket gave an average price that was more than four times higher than those who were not allowed to choose their own ticket. It appeared that those who chose their own tickets thought their chances of winning were higher and, therefore, their tickets were more valuable. The participants seemed to hold an illusory correlation between choosing their own ticket and their chances of winning.

Accuracy of Correlation Detection

We have focused on a number of imperfections in people's ability to detect correlations. However, people are not completely inept at assessing correlations. Because the ability to detect correlations in our everyday lives is an important skill, one would expect people to accurately assess some correlations. For example, you would have a serious problem if you could not detect the correlation between touching a hot stove burner and developing a burn on your hand. Thus, it seems likely that we would detect this correlation.

Some evidence suggests that, under certain circumstances, people are quite good at detecting correlations. Nisbett and Ross (1980) speculated that the conditions are sometimes ideal for the detection of correlations behind classically or operantly conditioned responses. For example, if you touch a hot stove burner, the negative consequence of pain or a burn is immediate, and this consequence occurs every time you touch the burner. Thus, in this situation the relationship between the two events—the behavior of touching the burner and the consequence of pain—is perfect, and the time interval between the events is extremely short. This may be an example of an ideal situation for the detection of correlations.

As a demonstration of classical conditioning in everyday life, I recently took a bottle of hot pepper sauce to my introductory psychology class and held it up for everyone to see. Many of the students who use hot sauces on their meals reported salivating at the sight of this bottle. Does this remind you of Pavlov's dog? It appears that people have little trouble detecting the correlation between the sight of hot pepper sauce and the effects of consuming the sauce. As with the stove burner example, the time between these two events is typically short, and the events are strongly correlated.

In other situations in which the correlations between events are strong and our prior expectations or theories do not bias our assessment of these correlations, our assessments of these relationships may be fairly accurate (e.g., Alloy & Tabachnik, 1984; Jennings, Amabile, & Ross, 1982; Nisbett & Ross, 1980). For example,

Jennings, Amabile, and Ross tested college students' ability to detect correlations of varying strength from "theory-free," or "expectation-free" data. For one of these tasks, they presented students with drawings of men of varying heights holding walking sticks of varying heights, and the students were asked to estimate the strength of the relationship between the heights of the men and their walking sticks. When the actual correlations between these two variables were weak or moderate, the students had difficulty accurately detecting them, but when the actual correlations were very strong, the students were much better at detecting them (although their strength estimates were still lower than the actual strength of the correlations).

Kunda and Nisbett (1986) suspected that people are likely to be fairly accurate at assessing everyday correlations when the correlation concerns an area familiar to them and when the variables are easy to code or score. For example, how likely is it that you and another student would agree on your evaluations of the same college course? Kunda and Nisbett predicted that students' estimates of the degree of agreement—or correlation—between students' course evaluations would closely match the actual correlation, because students are very familiar with one another's impressions of courses. As they predicted, the students' estimates were extremely close to the actual correlation. However, when Kunda and Nisbett asked people to estimate correlations in areas with which they were much less familiar, the participants did not perform as well. For example, when undergraduates were asked to estimate the degree of agreement between reviewers of manuscripts submitted to a professional journal for publication, they vastly overestimated this correlation. Finally, in areas with which students were familiar but the variables were difficult to code, they also had difficulty accurately estimating correlations. For example, when they estimated the consistency of people's social behaviors (e.g. honesty and friendliness) across situations, they vastly overestimated the actual degree of consistency. However, when they estimated the cross-situational consistency of behaviors that are more easily coded (e.g., basketball scoring ability), they were much more accurate.

Summary

Our discussion of the mistakes people make in assessing correlations should make you think twice when someone tells you about the everyday correlations they have seen with their own eyes. People see all sorts of relationships in their daily lives that do not actually exist. For example, people believe that others exhibit a highly consistent degree of friendliness and honesty across a variety of situations, but they do not. People believe that physically attractive people are more intelligent than those who are unattractive, but they are not. Illusory correlations like these might develop as a result of the use of incomplete data, a positive-test strategy for testing hypotheses, the influence of biased expectations on the way we process or interpret information, or the rarity or distinctiveness of information we have about various groups. These illusory correlations may form the basis for negative stereotypes about groups of people, or they may lure us into believing we have more control over the events in our lives than we actually do (i.e., an illusion of control). Although there are a number of ways that we might develop inaccurate beliefs about the relationships between the events we observe in our everyday lives, people are also capable of making fairly accurate assessments of correlations. They appear to be most accurate when the correlations they attempt to detect are strong ones, when their prior expectations or theories do not bias their assessment of the correlations, and when they are assessing correlations between codable variables in areas familiar to them.

● PRACTICE PROBLEMS

1. A psychiatric nurse believes that people with mental disorders are more likely than other types of patients to admit themselves to the hospital on evenings during which there is a full moon. She tells you that she has noticed a full moon on a number of evenings during which several people admitted themselves to the hospital because of mental problems. Does she have enough evidence to make an accurate conclusion about this correlation? If not, what other evidence does she need?

2. Redelmeier and Tversky (1996) indicated that there is a widespread belief among physicians and their patients that the weather influences arthritis pain. However, when they examined the actual relationship between weather conditions and patients' arthritis pain, they found no correlation. Explain one way that patients might have developed this illusory correlation.

3. Steve, a high school teacher, told his students that he believes that parents with higher levels of education are especially unlikely to discipline their children. He asked his students to test his hypothesis by interviewing highly educated parents in the area. He told the students to ask the parents for examples of times when they should have disciplined their children within the past week but failed to do so. His students came back with a lot of these examples, and Steve concluded, "It looks as though there is a lot of evidence here to support my hypothesis that highly educated parents are especially unlikely to discipline their children." This chapter discussed a number of difficulties people have in assessing correlations; which ones are related to the way Steve and his students tested his hypothesis?

4. Imagine that a student collected data to determine if athletes are more outgoing than nonathletes. She observed a number of athletes in their everyday social interactions with others and wrote down whether they seemed outgoing.

She made similar observations of nonathletes. Examine the data below and determine if there is a correlation between these two attributes.

| | Outgoing? | |
Athlete?	Yes	No
Yes	90	30
No	54	18

6

Regression and Prediction

> Many a quack has made a good living from regression toward the mean.
> *Campbell and Kenny (1999)*

Consider the following situation, presented in the Chapter 1 questionnaire:

5. You were a good student in high school and expect to do quite well in college. In your first college course, Introduction to Psychology, you decide to put forth your best effort. You read the textbook carefully every week, go to every class, and take copious notes. However, when it comes time to study for the first exam, you come down with the flu and are able to study only half as long as you would have liked. To your surprise, you obtain the highest possible score (100) on the exam. You feel elated, especially after being told that the class average was 72. You continue to work hard in the course. When it comes time to study for the second exam, you study long and hard, spending much more time than you did for the first exam. Nevertheless, your score on the second exam drops to 92 (again, the class average was 72). How would you explain this result? Should you conclude that you studied too much?

Although you might be tempted to conclude that the change in your study habits caused the lower grade, there is a perfectly plausible statistical explanation for the lower grade; it is called **regression toward the mean.** This means that extreme values on one variable tend to be accompanied by less extreme values on a second variable when the two variables are imperfectly correlated. Regression toward the mean is a very important phenomenon with wide-ranging implications for the predictions and conclusions we make in our everyday lives. However, it is a phenomenon that people—even professional researchers—often misunderstand or overlook (Campbell & Kenny, 1999).

This chapter will take a close look at the phenomenon of regression toward the mean. First, the relationship between the topics of correlation and regression will be examined. Then, we will look at how a simple regression equation might be used to make predictions and how this equation can be used to understand the phenomenon of regression toward the mean. Finally, we will discuss (a) how good we are at making predictions, (b) everyday examples of regression toward the mean, and (c) the relationship between the law of large numbers (LLN) and regression toward the mean.

Correlation, Prediction, and Regression

As discussed in Chapter 5, when two variables are correlated with one another, we can use one of the variables to predict the other. For example, if students' scores on their first exam are typically correlated with their scores on the second exam, we could use a student's score on the first exam to predict how well the student might perform on the second exam. The accuracy with which we can make such a prediction depends on the strength of the correlation between students' scores on the first and second exams. If this correlation were perfect ($r = 1$), then we could use students' first exam scores to make perfect predictions of their scores on the second exam. If this correlation were nonexistent ($r = 0$), then the students' scores on the first exam would not help us at all in our task of predicting their scores on the second exam.

The statistical procedure that researchers often use to predict events is linear regression, which involves finding the best-fitting straight line (called the regression line) for a set of data, and all of the predicted scores fall on that line. Figure 6.1 shows what the regression lines would look like for a perfect correlation ($r = 1$), no correlation ($r = 0$), and a fairly strong correlation ($r = .7$) between first and second exam scores. The first exam score is indicated on the X-axis and the second exam score is indicated on the Y-axis. To make our discussion easier, I selected only three first-exam scores—100, 72, and 44—to create this graph (note that 72 is the average score on each exam).

If you look at the regression line for a perfect correlation, you can see that a person who scored 100 on the first exam is predicted to have a score of 100 on the second exam, and a person who scored 44 on the first exam is predicted to have a score of 44 on the second exam. The regression line for the zero correlation shows that no matter what a student scored on the first exam, he or she is predicted to have an average score (72) on the second exam. As I mentioned above, if there is a zero correlation between first and second exam scores, then knowing the first exam score does not help us predict the second exam score. The best prediction we can make in this situation is to predict the mean (average) score, because the mean is often the most representative, or typical, score. Of course, our predictions in this case would not be very accurate. In other words, students' actual scores on the second exam would be scattered around the zero-correlation regression line: Some would be above it, some would be below it. In the case of a perfect correlation, however, students' actual scores fall directly on the regression line; in other words, there are no prediction errors.

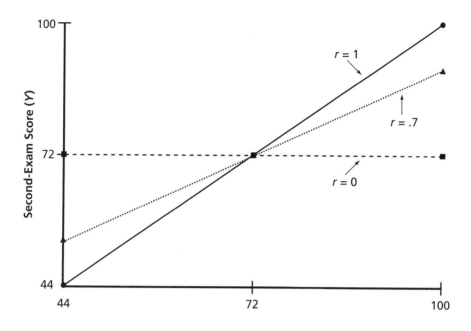

Figure 6.1 Regression lines for three different correlations

Correlations of $r = 1$ or $r = 0$ are rarely, if ever, found by researchers. A correlation of $r = .7$ is much more realistic. You can see in Figure 6.1 that the regression line for $r = .7$ falls in between the perfect-correlation and zero-correlation regression lines. This makes perfect sense, because $r = .7$ falls in between $r = 0$ and $r = 1$. We will discuss the regression line for $r = .7$ in greater detail below; first, I want to show you the simple regression equation used to create these lines, because this equation makes it much easier to understand the phenomenon of regression toward the mean.

The regression equation used to create Figure 6.1 is $\hat{Z}_y = rZ_x$, where r is the correlation coefficient, Z_x is the student's standardized score on the first exam, and $\hat{Z}_y$ is the predicted standardized score on the second exam.[1] A standardized score, or Z-score, reflects how many standard deviations a raw score is from its mean. For example, a Z-score of 0 indicates that the raw score is equal to the mean (i.e., it is 0 standard deviations away from the mean). A Z-score of 1 indicates that the raw score is one standard deviation above the mean, and a Z-score of −2 indicates that the raw

---

[1] I chose the regression equation for Z-scores (instead of raw scores) for a couple of reasons. First, regression toward the mean occurs when the predicted score is less extreme *in standardized units* than is the predictor score. Also, the Z-score version of the regression equation makes it easier to explain the relationship between the correlation coefficient and the degree of regression toward the mean (as discussed later).

score is two standard deviations below the mean. The formula for calculating Z-scores from raw scores is

$$Z = \frac{\text{score} - \text{mean}}{\text{standard deviation}}$$

For example, if you assume that the standard deviation of students' first-exam scores is 14, the Z-score that corresponds to a first-exam score of 72 is

$$Z = \frac{72 - 72}{14} = 0$$

This makes sense because 72 is the mean of the first-exam scores. A first-exam score of 100 translates into a Z-score of 2 as follows:

$$Z = \frac{100 - 72}{14} = \frac{28}{14} = 2$$

Thus, a score of 100 is 2 standard deviations, or 28 points, above the mean. Similarly, a score of 44 is 2 standard deviations, or 28 points, below the mean.

Figure 6.2 shows the same regression lines that were in Figure 6.1, but this time the exam scores have been translated into Z-scores. To give you a better feel for how the regression equation, $\hat{Z}_y = rZ_x$, relates to Figure 6.2, let's use it to calculate a few

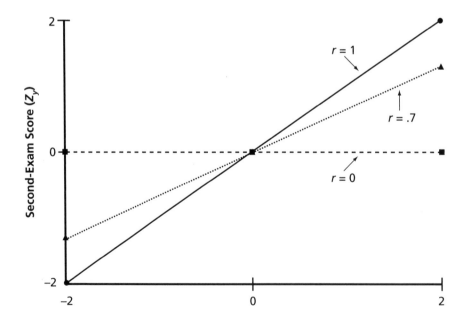

Figure 6.2 Regression lines for standardized exam scores

predicted scores. You can see that, when there is a perfect correlation ($r = 1$) and Z_x is 2 (i.e., the student obtained a score of 100 on the first exam), the predicted value for the second exam— $\hat{Z}_y$ —is also 2 (which corresponds to a raw score of 100)—that is, $\hat{Z}_y = rZ_x = 1(2) = 2$. Because $r = 1$, the predicted Z-score will always perfectly match the value of Z_x. However, when there is no correlation between students' first and second exam scores, $\hat{Z}_y$ will always equal 0 (i.e., the predicted score will match the mean score). Thus, if a student has a perfect score on the first exam ($Z_x = 2$), the predicted Z-score for the second exam is 0 [i.e., $\hat{Z}_y = rZ_x = 0(2) = 0$], which equals a raw score of 72.

Regression Toward the Mean

An interesting thing happens when you consider the regression line for $r = .7$ in Figure 6.2. When a student's first-exam Z-score is 2 (a perfect score), the predicted Z-score for the second exam is 1.4 [i.e., $\hat{Z}_y = rZ_x = .7(2) = 1.4$]. Note that this predicted score is closer to the mean of $Z_y(0)$ than the first-exam Z-score of 2 is to the mean of $Z_x(0)$. In fact, it is .6 points closer to its mean (i.e., $2.0 - 1.4 = .6$). In other words, the predicted score *regressed* toward the mean.

If we translate the predicted Z-score of 1.4 into a predicted raw second-exam score (Y) using a rearranged version of the Z-score formula, we see that the predicted score is 91.6 [i.e., $\hat{Y}$ = mean + $\hat{Z}_y$(standard deviation) = 72 + 1.4(14) = 91.6].[2] Thus, a person who received a score of 100 on the first exam is predicted to have a score of about 92 on the second exam. Such predictions are rarely perfect, but note that these are the same two scores that you might have found so puzzling in the example at the beginning of this chapter. The drop in the student's performance from the first exam to the second exam can be explained by the phenomenon of regression toward the mean; we do not need to invent other explanations for this drop.

Although we had to go through some rather boring formulas to get this insight, I hope you find the insight itself absolutely fascinating. Before you assume that this fascination would make you a geek, consider this: If you experienced such a drop in your exam scores and attributed it to too much studying or some other cause, you might have made some ill-advised changes in your behavior (e.g., studying less) that would have resulted in even more serious problems (e.g., flunking the course). Moreover, this is just the tip of the iceberg. As implied earlier, regression toward the mean is a pervasive phenomenon that could lead us to make a number of inaccurate conclusions in our daily lives. Before examining a wider variety of everyday examples, let's finish our important discussion of the regression equation and how it relates to the phenomenon of regression toward the mean.

Notice that regression toward the mean also occurs when we consider extreme first-exam scores that are below the mean. Once again, take a look at the regression

---

[2] I made the assumption that the standard deviation of Y is 14, the same value used for the standard deviation of X.

line for $r = .7$. When a student's first-exam Z-score is -2, the predicted Z-score for the second exam is -1.4 [i.e., $\hat{Z}_y = rZ_x = .7(-2) = -1.4$]. This predicted score is closer to the mean of $Z_y(0)$ than the first-exam Z-score of -2 is to the mean of $Z_x(0)$. If we translate these Z-scores into raw scores, we can see that a student who scored a 44 ($Z_x = -2$) on the first exam is predicted to score a 52.4 ($\hat{Z}_y = -1.4$) on the second exam. That is an increase of 8.4 points caused solely by regression toward the mean. Imagine that you obtained a score of 44 on the first exam and, as a result, decided to study with a friend for the next exam. If your score were 8 points higher on the second exam, would you be tempted to conclude that studying with a friend helped your performance? It is an awfully tempting conclusion, but it could be completely incorrect. Studying with a friend is one possible explanation for the improvement, but so is regression toward the mean.

So far, we have looked at only the very extreme first-exam scores of 44 and 100. Regression toward the mean is more likely to occur with such extremes scores than it is with less extreme scores. If you look at the regression line for $r = .7$, you can see that this line gets closer and closer to the perfect-correlation line ($r = 1$) as you move closer to the mean. In other words, there is less regression toward the mean as you move closer to the mean. For example, a person with a first-exam Z-score of 1 would have a predicted second-exam Z-score of .7 [i.e., $\hat{Z}_y = rZ_x = .7(1) = .7$]. This represents a regression toward the mean of only .3 points (i.e., $1 - .7 = .3$) as compared with the regression of .6 points we saw earlier for a first-exam Z-score of 2. Moreover, for a person whose first-exam score falls on the mean ($Z_x = 0$), there is no regression toward the mean at all; this person's predicted second-exam Z-score is also 0.

We have discussed the fact that, solely because of regression toward the mean, a student who scored a 100 on the first exam might have his or her score drop to 92 on the second exam, and a student who scored a 44 on the first exam might have his or her score increase to 52 on the second exam. You might be wondering why this happens. Why does regression toward the mean occur? What causes it? The answer is quite simple: a less-than-perfect correlation between two variables. That's it; no mysterious forces or supernatural explanations are needed to account for the phenomenon of regression toward the mean. To explain why this happens with students' exam scores, for example, we can simply say that students' scores on the first exam in a class are not perfectly correlated with their scores on the second exam. There are a number of reasons for this imperfect relationship. If you think about why your score on one exam may be different from your score on another exam, you could probably come up with a number of reasons, including (a) the questions on one exam overlapped with your knowledge more closely than did the questions on the other exam, (b) you slept unusually well the night before one of the exams, and (c) you made more lucky guesses on one of the exams. If you obtained a near-perfect score on an exam, it may be partly due to the fact that all of these conditions were in your favor at the time. The conditions are likely to be less ideal for the second exam, and your score is likely to be lower. Similarly, if you obtained a horrible score on one of the exams, it may be partly due to the fact that all of these conditions were working against you at the time. Those conditions are likely to be better for the second exam, and your score is likely to improve. Thus, even if your true academic ability remains unchanged from one exam to the next, random or chance factors can affect your scores, creating some inconsistency in your performance from one exam to the next (i.e., your performances from one exam to the next will be imperfectly correlated).

At this point, you might be wondering if it is ever possible for a student to get the high score on two different exams. Of course it is! You may have done this yourself. Regression toward the mean does not apply to every student's *actual* scores; however, it does apply to every student's *predicted* scores. It is certainly possible that a student who obtained a perfect score on the first exam will also obtain a perfect score on the second exam. But, if we were to look at the *average* of the second-exam scores for *everyone* who scored 100 on the first exam (or $Z_x = 2$), we would see regression toward the mean. In other words, on the average, people who scored extremely well on the first exam will score lower on the second exam.

If you look back at the regression equation, you can see that any time we have a less-than-perfect correlation ($r < 1$) the predicted score ($\hat{Z}_y$) has to be less extreme than the predictor (Z_x). As we saw above, if $r = .7$ and Z_x is 2, then the predicted score is 1.4. Another important aspect of regression toward the mean that is illustrated by this regression equation is that weaker correlations result in more regression toward the mean. For instance, if $Z_x = 2$ and $r = .4$, then the predicted score is .8 (not 1.4, as it was when $r = .7$). This reflects a regression toward the mean of 1.2 points (i.e., 2 − .8 = 1.2) instead of the .6 points discussed earlier. If $Z_x = 2$ and $r = .2$, then the predicted score is .4, which is very close to the mean of 0. This all makes sense when you consider the fact that we can make better predictions when there is a stronger correlation between two variables. When the correlation is perfect, we can make perfect predictions that are far from the mean. As the correlation becomes less perfect, we have to rely less on our predictor variable and move our predictions closer to the mean. In the extreme case of a zero correlation, our predictions should match the mean.

How Good Are We at Making Predictions?

Now that we have reviewed the basic statistical principles behind linear regression, we can take a look at how well people seem to follow these principles when they make intuitive predictions. Of course, people do not carry around regression equations and calculators in order to make predictions in their everyday lives. Thus, it would not be surprising if their predictions were not exactly like those generated by regression equations. Nevertheless, we might expect that people who are very knowledgeable or experienced in a particular area are able to generate accurate predictions within their area of expertise. For example, we might expect accurate predictions from (a) clinical psychologists predicting patient outcomes (e.g., risk of future violent behavior), (b) physicians predicting how long someone with a fatal disease has to live, (c) admissions committee members predicting the performance of incoming graduate students, (d) personnel managers predicting the performance of job applicants, and (e) stockbrokers predicting the growth of companies. However, research that has compared people's intuitive (or "clinical") predictions with statistical predictions (e.g., generated by regression formulas) has found that the human judges are often outperformed by the statistical predictions (e.g., Dawes, Faust, & Meehl, 1989; Meehl, 1954; Nisbett & Ross, 1980). This occurs even when the human judges are experts in their field and they have the same information that was used to make the statistical predictions.

Humans' intuitive predictions are not as accurate as statistical predictions for a number of reasons, including the facts that we have a tendency to form illusory correlations (as mentioned in Chapter 5) and we often fail to use our own prediction rules consistently from one situation to the next (see Dawes et al., 1989). Another reason—one that is very relevant to this chapter—is that people's predictions tend not to regress toward the mean as much as they should. Before we consider some classic research on this problem, please answer the following questions:

1. At the end of his freshman year, a student's grade-point average was better than that achieved by 85% of his class (i.e., his percentile score was 85). Assuming grade-point averages range from 0 to 4.0, what would you predict his grade-point average was for that year? _____ What grade-point average would you predict for a student whose percentile score was 10%? _____

2. A group of students took a test that measured their ability to concentrate and to extract information from complex messages. Students who score high on this test tend to have high grade-point averages, but performance on the test can also be influenced by a person's mood and mental state (e.g., sleepiness) at the time of the test. If a student took this test and scored better than 85% of the other students who took the test, what would you predict this student's grade-point average to be? _____ What grade-point average would you predict for a student who scored better than only 10% of the other students on this test? _____

3. A number of students were given a test of their sense of humor. In general, students who score high on this test tend to have higher grade-point averages than those who score low, but it is not possible to predict grade-point average from sense of humor with great accuracy. If a student took this sense of humor test and scored better than 85% of the other students who took the test, what would you predict this student's grade-point average to be? _____ What grade-point average would you predict for a student who scored better than only 10% of the other students? _____

For each of these three scenarios, what grade-point average did you predict for the student who scored at the 85th percentile? Are these three predicted grade-point averages very similar to each other, or do they decrease as you move from Question 1 to Question 3? Do you see how these scenarios are related to our earlier discussion of the effect of the strength of a correlation on the degree of regression toward the mean? In the first scenario, we have a predictor variable (grade-point average translated into a percentile) that should be highly correlated with students' grade-point averages. Thus, we should expect a student who is at the 85th percentile to have a grade-point average that is relatively high. In the second scenario, however, our predictor variable is not as highly correlated with grade-point average. In this case, our prediction of the student's grade-point average should regress toward the mean a bit. In other words, if you predicted a grade-point average of 3.4 for the student in the first scenario, you should have predicted a grade-point average a bit lower for the student in the second scenario. Because the predictor variable in the third scenario—sense of humor—should be correlated even less well with grade-point averages, your prediction for this student should be lower than your other two predictions (i.e., more regression toward the mean would be expected because of the lower correlation).

What grade-point averages did you predict for the students who scored at the 10th percentile? Are these three predicted grade-point averages very similar to each other, or do they *increase* as you move from Question 1 to Question 3? Once again, more regression toward the mean is expected when the predictor variable has a lower correlation with grade-point averages. Thus, your predictions should regress more and more toward the mean—resulting in *higher* predictions—as you move from Question 1 to Question 3.

These three scenarios are similar to those used by Kahneman and Tversky (1973) in their research on people's intuitive predictions. They gave each of three separate groups of participants one of the three scenarios and asked them to predict the grade-point averages. Unlike my questions, however, they gave their participants more than just two predictor scores (i.e., percentiles). Their participants were asked to estimate grade-point averages for 10 different hypothetical students who had a range of percentile scores on the predictor variable. They found that the participants who predicted grade-point averages using students' percentile scores on the test of mental concentration did not make predictions that regressed toward the mean. Their predictions closely matched the students' scores on the predictor variable; target persons with very high mental concentration scores were predicted to have very high grade-point averages, and target persons with very low mental concentration scores were predicted to have very low grade-point averages. Kahneman and Tversky suggested that this is another example of the use of the representativeness heuristic. In other words, their participants seemed to be predicting the outcome that was most similar to—or representative of—the scores on the predictor variable. For example, a very high grade-point average seems representative of a person who had a very high score on a test of mental concentration. This strategy, however, results in predictions that do not regress toward the mean. The participants who predicted grade-point averages using students' percentile scores on the test of sense of humor made predictions that were quite similar to the predictions made by the other two groups of participants, but they did exhibit a little more regression toward the mean than did the other two groups' predictions. Thus, these participants seemed to realize that their predictor variable was not a perfect predictor of grade-point average.

Regression Toward the Mean in Everyday Life

Kahneman and Tversky (1973) noted that it might be very difficult for people to understand and apply the concept of regression toward the mean because it violates our notion that a predicted outcome should be similar to, or representative of, the value of the predictor. For instance, we expect highly intelligent people to exhibit outstanding achievement, extremely talented athletes to put forth extremely talented performances, and very tall parents to have very tall children. Of course, if two variables are positively correlated, we should expect higher values on one variable (e.g., father's height) to correspond with higher values on the second variable (e.g., son's height). The problem is that our predictions do not adequately take into account the fact that extreme values on one variable tend be matched by less extreme—not equally extreme—values on the second variable.

The difficulty we have in recognizing the phenomenon of regression toward the mean can lead us to make inaccurate conclusions about events in our everyday lives. When regression toward the mean happens and we notice that an outcome does not match our nonregressive prediction, we often attribute it to some intuitively appealing cause other than regression toward the mean. Let's take a look at some interesting examples.

Examples from the World of Sports

In January, 2000, I watched with great interest as a nonregressive prediction by a well-known television personality backfired. Early in the 1999 National Football League (NFL) season, Mike Wallace—a reporter for *60 Minutes*—planned an interview with Bill Parcells, who was the head coach of the New York Jets. After the Jets had a great 1998 season, achieving a win-loss record of 12-4, Wallace expected the Jets to be bound for the playoffs once again at the end of the 1999 season. The controversial coach appeared to be poised to take yet another team to the Super Bowl. It would have made an interesting story, but it did not happen. The Jets won only 8 games in 1999, but Mike Wallace did the interview anyway.

During the segment of the *60 Minutes* show that contained his interview, Mr. Wallace admitted that they made a faulty prediction, and they were surprised that the Jets had a relatively poor season that year. Given our earlier discussion of how people use the representativeness heuristic to make predictions, I suspect that Mr. Wallace expected virtually no regression toward the mean for the Jets' 1999 season. The Jets had an excellent record in 1998, and an excellent 1999 season *seems* much more representative of such a team than does a mediocre season. However, because NFL teams' records from year to year are imperfectly correlated, the teams that have outstanding records one year will have less outstanding records—on the average—the following year. Thus, it is not surprising that the Jets' 1999 season was not as good as their 1998 season.[3]

Because of regression toward the mean, NFL teams that have horrendous records one year will have less horrendous records—on the average—the following year. A popular response to a very poor season for a NFL team seems to be firing the head coach. When the team shows improvement the following year, that improvement will likely be attributed to the new coach rather than regression toward the mean. For example, after the 1999 NFL season, the *Cincinnati Enquirer* ("Around the NFL," 2000) reported that "new was better." The report stated, "The eight teams with new coaches . . . collectively improved significantly." Of course, most of those teams had worse-than-average records the previous year. Thus, the improvements would have occurred—on the average—even without the coaching changes. Notice how chance improvements in team performances after coaching changes might lead to an illusion of control (see p. 70 for a brief discussion of illusion of control). Even when the coaching change has no impact on the team's performance, the team's management will probably believe that it did; thus, they might come to believe—incorrectly—that they had some control over the team's performance.

We have discussed how regression toward the mean relates to team performances from year to year. Individual players' performances are similarly affected by regression toward the mean. Nisbett and Ross (1980) explained that the phenomenon known as the "sophomore slump" among baseball players is probably a result of regression toward the mean. The sophomore slump occurs when an outstanding rookie player has a more mediocre performance during his second, or sophomore, year. Of course, because players' performances are imperfectly correlated from year to year, regression toward the mean dictates that outstanding first-year players will—on the average—perform less well in their second year. Although sports fans may attribute the decline to the first-year fame spoiling the player, the decline is probably a result of an imperfect correlation between players' year-to-year performances.

---

[3]It is an interesting coincidence that the Jets' fall from 12 wins in 1998 to 8 wins in 1999 is exactly what would be predicted by a regression equation. The correlation between NFL teams' wins in 1998 and 1999 is extremely close to zero ($r = -.01$). As we saw earlier, when two variables have a zero correlation, the best prediction one can make is the mean. Because the average number of wins for NFL teams in 1999 was 8, the Jets' 8 wins perfectly match this prediction (i.e., the mean). However, because there is no correlation between teams' wins in 1998 and 1999, wins in 1998 is a worthless predictor of wins in 1999; one would end up predicting 8 wins for *all* NFL teams in 1999.

I should point out that the correlation of $r = -.01$ between NFL teams' wins in 1998 and 1999 appears to be unusually low. The year-to-year correlations for the 1995–2000 NFL seasons range from $r = -.01$ to $r = .50$, averaging $r = .32$. If a person expected the 1998–1999 correlation to be higher (e.g., $r = .32$) than it actually was, less regression toward the mean would be predicted for the 1999 season. Nevertheless, $r = .32$ is not a very strong correlation, and one would still expect a substantial degree of regression toward the mean with such a correlation.

Ozone and Gas Prices

Possible misinterpretations of phenomena that reflect regression toward the mean are not difficult to find in daily news stories. In fact, the day that I wrote this section of the book, such a story appeared on the front page of the *Cincinnati Enquirer* (Pilcher, 2000). Gas prices that summer had increased substantially (i.e., as much as 70%) compared with the previous summer. The negative effects of such an increase had been the focus of many media reports. However, Pilcher thought he had spotted a potentially positive effect of the increase in gas prices. He wrote, "And now, some good news about increased gas prices: They may be helping to increase area bus ridership and clean the Tristate's air" (p. A1).

Pilcher (2000) presented some data that showed that ozone levels at various monitoring sites around Cincinnati had dropped in June of 2000 compared with the previous year. Was the drop a result of the increase in gas prices? It seems logical that if the increase in gas prices led to fewer cars on the road, then lower ozone levels might result. But, an important clue that suggests another cause for the drop in ozone levels was contained in a paragraph near the end of the article: "The Tristate's ozone levels, meanwhile, declined in June compared with last year, when many of the city's seven monitoring sites recorded *five-year highs* [emphasis added]" (p. A9).

Clearly, Cincinnati's ozone levels were extreme in the summer of 1999. Because ozone levels are imperfectly correlated from year to year, one would expect a drop in ozone levels in the summer of 2000 because of regression toward the mean. To his credit, Pilcher acknowledged that his explanation for the drop in ozone levels was unproven, and I must admit that I cannot prove it was due solely to regression toward the mean. Nevertheless, it appears very likely that the drop was at least partly a result of regression toward the mean, and this explanation was never mentioned in the article—not even by an environmental engineering professor who was quoted in the article.

The Magazine Curse

Kruger, Savitsky, and Gilovich (1999) pointed out that regression toward the mean might account for the decline in actors' or athletes' success that often follows their appearance in popular magazines. They explained, for example, that some athletes are afraid to be featured on the cover of *Sports Illustrated* because they fear that it will jinx them, causing a decline in their athletic performance afterward. Interestingly, an appearance on the cover of *Sports Illustrated* probably *would* be followed by a decline in performance. But, the decline would not be caused by the appearance in the magazine; it would be caused by regression toward the mean. Athletes are probably most likely to be featured on the cover of *Sports Illustrated* when they have accomplished outstanding athletic feats. As discussed earlier, athletes' performances from year to year (or month to month) are imperfectly correlated. Thus, one would expect that an outstanding performance—such as the type that would get an athlete featured on the cover of *Sports Illustrated*—would be followed by a more mediocre performance after the publication of the magazine.

A similar effect might occur among actors. Actors are probably most likely to be featured prominently in a popular magazine when they are experiencing tremendous success and fame as an actor. Because the popularity of actors at one point in time is likely to be imperfectly correlated with their popularity at another point in time, we would expect that tremendous popularity—such as the type that would get actors featured in a popular magazine—would be followed by a lower level of popularity after the publication of the story.

Alternative Health Remedies

Imagine that you have felt quite depressed lately, and you decide it is time to try a new herbal remedy that is supposed to improve mood. After taking it daily for several weeks, you feel less depressed. Would you conclude that the herbal remedy worked? Unfortunately, people sometimes rely on this type of evidence to conclude that a remedy or treatment is effective; but such a conclusion is horribly flawed. One major flaw is that the improvement may be due to the well-known placebo effect: If we believe a treatment will be effective, this belief may cause an improvement in our condition. A second major flaw is that the improvement may be due to regression toward the mean.

Kruger et al. (1999) and Campbell and Kenny (1999) explained that regression toward the mean is relevant to changes in people's mental and physical health over time, because we are most likely to seek treatment when our condition is at its worst. For example, people may be most likely to seek treatment for depression when the depression is at its worst. Because people's degree of depression at one point in time is imperfectly correlated with their degree of depression at a second point in time, we can expect regression toward the mean. In other words, if you seek treatment for your depression because it is particularly severe, your depression is likely to improve after "treatment" simply because of regression toward the mean. In fact, Campbell and Kenny (1999) discussed research evidence that demonstrates that depressed people who seek therapy become much less depressed over time even if they do not actually receive any therapy.

Similar "regressive" improvements in people's mental or physical health may happen to coincide with their efforts to improve their health with a variety of alternative treatments, from herbal remedies to subliminal tapes. But it is a mistake to conclude that the treatments caused the improvement. Perhaps now you can better understand the quote at the beginning of this chapter: "Many a quack has made a good living from regression toward the mean" (Campbell & Kenny, 1999, p. 48).

Other Ways to Conceptualize Regression Toward the Mean

We have seen that regression toward the mean can be found in a variety of events in our everyday lives. In fact, regression toward the mean can occur whenever any two variables—which I will refer to as X and Y—are imperfectly correlated. There are several different ways of thinking about regression toward the mean, depending on what

X and Y represent. So far, we have discussed regression toward the mean for situations in which X and Y represent the same variable measured at different points in time (e.g., an athlete's performance at two different points in time). However, regression toward the mean also occurs when X and Y represent the same variable measured on two different people. Perhaps the most famous example of this is Sir Francis Galton's (1886) discovery that, although tall parents tend to have tall children, the adult children are not quite as tall as their parents, and short parents tend to have children that are not quite as short. At first, Galton suspected there must be some biological force at work that caused offspring to regress toward the mean. But, later he realized that this explanation could not be correct because regression toward the mean also works backward through time: Tall children tend to have parents that are not quite as tall, and short children tend to have parents that are not quite as short. Regression toward the mean works in both directions; extreme values on variable X tend to be accompanied by less extreme values on variable Y, and extreme values on variable Y tend to be accompanied by less extreme values on variable X. Campbell and Kenny (1999) extended this example to characteristics of parents and children other than height. For example, we should find that (a) very wealthy parents tend to have less wealthy children, (b) very intelligent parents tend to have less intelligent children, and (c) very creative parents tend to have less creative children.

Finally, regression toward the mean also occurs when X and Y represent two different qualities of the same person measured at one point in time. For example, a person who is an outstanding golfer might be relatively less adept at tennis, and a person who is an outstanding researcher might be relatively less adept at teaching.

The Law of Large Numbers and Regression Toward the Mean

Please answer the following question posed by Nisbett, Krantz, Jepson, and Kunda (1983) in their study of everyday reasoning:

> Harold is the coach for a high school football team. One of his jobs is selecting new members of the varsity team. He says the following of his experience: "Every year we add 10–20 younger boys to the team on the basis of their performance at the try-out practice. Usually the staff and I are extremely excited about the potential of two or three of these kids—one who throws several brilliant passes or another who kicks several field goals from a remarkable distance. Unfortunately, most of these kids turn out to be only somewhat better than the rest." Why do you suppose that the coach usually has to revise downward his opinion of players that he originally thought were brilliant? (p. 354)

Perhaps by now you were able to quickly recognize this as an example of regression toward the mean: an extremely good performance followed by a more mediocre performance. But did you also recognize that this example is related to the law of large numbers?

As discussed in Chapter 3, the law of large numbers dictates that larger samples will provide us with better estimates of population values than will smaller samples. Note that Harold's initial impression of the two or three outstanding football players was based on a small sample of their behavior: one try-out practice. As shown in Chapter 3, a small sample is more likely than a large sample to leave us with an inaccurate, or extreme, impression of the "population" of a person's behavior. For example, you can probably imagine an otherwise average golfer making one outstanding shot or an average student performing unusually well on one quiz. Similarly, the two or three players who performed extremely well at the try-out practice might have had atypical performances. The players' true athletic ability can be estimated more accurately from a larger sample of their athletic performances—for example, their performances over the course of an entire football season. Thus, once we have the larger sample, the player's ability will appear to be less outstanding (i.e., it will regress toward the mean).

We could apply the same reasoning to the impressions we form of the quality of food at restaurants. If you go a restaurant and order one item from the menu, you might select the only menu item that is outstanding. If you go back to the restaurant several times and try more and more menu items—thereby obtaining a larger and larger sample—you will begin to view the food as less outstanding. Imagine going to a restaurant and sampling the entire menu on your first visit and sampling the entire menu once again on your second visit (let's assume that you can consume a lot of food without vomiting). If we assume that the actual quality of food was consistent from the first visit to the second visit, your impressions of the quality of the restaurant's food at these two points in time are likely to be very similar to each other; after all, you ate everything! In this situation, regression toward the mean would be virtually nonexistent.

These examples illustrate that when the value of one variable (X) is based on a small sample and the value of another variable (Y) is based on a large sample, the two variables will probably be imperfectly correlated, and thus, regression toward the mean will occur. However, when the values X and Y are both based on large samples, those variables will be more highly correlated. As we have seen, when two variables are highly correlated, there is less likely to be regression toward the mean.

Summary

When two variables are correlated with one another, we can use one of the variables to predict the other. However, when the two variables are imperfectly correlated, extreme values on one variable tend to be accompanied by less extreme values on a second variable. This phenomenon is called regression toward the mean. Regression toward the mean pervades many aspects of our everyday lives. For instance, it can be found in the tendency for (a) athletes to have a relatively mediocre season after having a stellar one, (b) actors to experience a lower degree of success after being featured in a popular magazine, (c) our mental or physical health to improve after we try alternative health remedies, and (d) tall parents to have shorter children. Although we witness regression toward the mean in our daily lives, we often fail to recognize it for what it is. Instead, we invent inaccurate or superfluous causal explanations for the changes we observe, attributing improved health to an alternative treatment or decreased athletic success to a decline in motivation. Finally, regression toward the mean can be a result of an inadequate sample size for one or both of the variables of interest. For example, if we have a small sample for our predictor variable (e.g., an athlete's performance during a brief try-out) and a large sample for our outcome variable (e.g., an athlete's performance during an entire season), we will see regression toward the mean because of the change in sample size—especially when the small sample contains extreme values.

● PRACTICE PROBLEMS

1. You are the agent for one of the top professional female basketball players in the country. She has had an outstanding year and led the league in points scored. Although her contract expires next year and she will probably be sought by a number of other teams, she wants you to negotiate a better contract with her current team. You do so, and the team agrees to give her a five-year contract that will make her the highest-paid female player in the history of professional basketball. However, the following year her performance declines. At least seven other women in the league (all of whom are paid less money) score more points than she does. Why did this happen?

2. Kahneman and Tversky (1973) stated that regression toward the mean might lead people to conclude that punishment is more effective for modifying a person's behavior than are rewards. Explain why this might be true. (*Hint:* Think of the types of behaviors that typically precede rewards and punishments.)

3. Kruger et al. (1999) indicated that many of us believe that drawing attention to our good fortune or success invites disaster. For instance, golfers who are doing extremely well on the first portion of the golf course might avoid adding up their score at that point—or even mentioning their success to their golf partners—for fear that doing so will jinx their performance on the rest of the course. Similarly, vacationers might not want to mention that the weather has been perfect during the first few days of their vacation, because they fear that doing so will surely bring bad weather. Describe how regression toward the mean might contribute to the development of such beliefs.

4. Two high schools gave their juniors a practice Scholastic Aptitude Test (SAT). One school encouraged all of these students to take a course designed to improve their SAT scores on the actual test, and the second school gave the same advice to only those students who performed poorly on the practice test. After these students took the course and the actual SAT, a comparison was made of their practice test scores and their actual SAT scores. As it turned out, the students at the second school showed more improvement after taking the course than did the students at the first school. Should we conclude that the SAT course had a greater impact on the students at the second school? Why?

Factorial Analysis of Variance and Attribution

> Although people sometimes acknowledge the existence of multiple causes, it is clear that they frequently act in ways far more consistent with beliefs in unitary causation.
>
> *Nisbett and Ross (1980)*

Recall Question 6 from the questionnaire in Chapter 1:

6. Imagine that you volunteered to participate in a psychological study. When you arrive for the experiment, you are ushered to a small room. You put on headphones and listen to tape-recorded instructions that explain that you will be participating in a discussion with five other students about personal problems faced by college students.... [During the discussion the first student explains] that he needs help because he is having a seizure. He continues to beg for help, then starts choking while saying that he is going to die. You hear him choke again, and then the intercom system cuts off his voice. What would you do in this situation? Would you leave your room and try to help this desperate student? If I told you that a person was actually in this situation and he did not help, what would you think of him? Would you view him differently if I told you that his response was fairly typical of other people who were in this situation?

In a moment, we will consider how your answers to these questions relate to the topic of factorial analysis of variance. First, we will discuss some of the basic concepts related to this statistical procedure. Afterward, we will consider how these concepts are useful for thinking about the causes of the events and behaviors we observe in our daily lives.

What Is a Factorial Analysis of Variance?

Analysis of variance (ANOVA) is a statistical procedure often used by researchers to evaluate differences among two or more groups of people. One common use of ANOVA is to determine the causal impact of one variable (the independent variable) on a second variable (the dependent variable). For instance, imagine that a researcher wants to investigate the effects of practice on golfers' performance. She has one group of golfers practice 1 hour a week and another group practice 5 hours a week; both groups do this for a month. Afterward, the golfers play several rounds of golf and record their scores. The researcher could compare the performance of the two groups of participants to determine the impact of the independent variable (the amount of practice) on the dependent variable (the golfers' scores).

A *factorial* ANOVA is used to analyze the results of experiments that have more than one independent variable. For instance, the same researcher might also want to investigate the effect of golf lessons—that is, instruction on the proper way to swing a golf club—on golfers' performance. She might have some of her participants receive an hour of instruction each week during the experiment, and other participants would receive no instruction. She could investigate the effects of the practice *and* instruction in the same study by creating four groups of participants. As shown in Table 7.1, depending on what group they were in, the participants would practice (a) 1 hour a week without instruction, (b) 5 hours a week without instruction, (c) 1 hour a week with instruction, or (d) 5 hours a week with instruction. This type of design is called a two-factor design because it has two independent variables. If a researcher examined only one independent variable in an experiment, it would be called a single-factor design.

One of the advantages of a two-factor design over a single-factor design is that the researcher can examine the effects of more than one independent variable in the same study. The effect of one independent variable in a factorial design is called a **main effect.** For example, the effect of practice on golfers' scores could be examined by comparing everyone who practiced 1 hour per week with everyone who practiced 5 hours per week; this would reflect the main effect of practice. Similarly, we could examine the main effect of instruction by comparing those who received the instruction with those who received no instruction. The researcher might find that both of these independent variables had main effects on the golfers' scores: Increased practice time resulted in better performance, and instruction resulted in better performance as well.

Another major advantage of a two-factor design over a single-factor design is that the researcher can examine the **interaction** between independent variables. An interaction occurs when the combination of the two independent variables produces ef-

Table 7.1

A Two-Factor Experiment to Examine the Effects of Practice and Instruction on Golfers' Performance

	Amount of Practice per Week	
Instruction?	1 Hour	5 Hours
No	Group A	Group B
Yes	Group C	Group D

fects that neither independent variable produces alone. In other words, the effect of one independent variable changes when the second independent variable enters the scene. For example, we might find that golfers who practiced 5 hours per week performed better than those who practiced 1 hour per week, but the effect of practice was especially pronounced among golfers who received instruction. This might occur because those who received instruction were more likely to be practicing a proper swing and were therefore more likely to improve as a result of practice. Thus, the effect of practice depended on whether the golfers had instruction.

How Is a Factorial ANOVA Related to Everyday Reasoning?

As stated earlier, researchers use a factorial ANOVA to determine the causes behind various events or behaviors. Let's take a look at how factorial ANOVA concepts can help us understand causality in our everyday lives.

Single Versus Multiple Causes

One important concept illustrated—at least implicitly—by a factorial ANOVA is that events and behaviors often have multiple causes. A factorial ANOVA, by definition, involves examining the effects of more than one independent variable on some outcome of interest. It is useful to keep in mind that the events and behaviors we observe in our everyday lives often have more than one cause. For instance, as suggested earlier, a golfer's performance might be affected by both practice and instruction. Of course, many more variables probably affect a golfer's performance, such as wind, course difficulty, fatigue, and so on. Although this example may seem obvious to you, Nisbett and Ross (1980) and Stanovich (2001) pointed out that we sometimes tend to focus only on one cause for an event, overlooking other possible causes. For instance, imagine reading a front-page newspaper article with the headline "The Cause of Depression Found." The article describes a research study in which a group of biologists discovered a genetic abnormality in a group of depressed patients. The author of the article discusses the potential benefits of this amazing "breakthrough" and suggests that advances in gene therapy might one day make it possible to completely eradicate depression in human beings.

Although the depression article is fictional, you may have seen similar articles that suggest single causes for psychological disorders or other behaviors. In fact, at the time I was writing this chapter, I read a newspaper article that blamed alcohol abuse for an increase in violent crime in Britain. The article said, "Violent crime in Britain has risen 16 percent over the last year . . . and the country's top law enforcement official blamed the increase on alcohol abuse" ("British crime rise," p. A16). The fact is that problems such as depression and violent crime have multiple causes. Genes may play a role in the formation of depression, but they are not the *only* cause of depression. Alcohol abuse may have contributed to Britain's increase in violent crime, but it is probably not the *only* cause of the increase in crime.

Biological, psychological, and social factors may all contribute to psychological or social problems, such as depression and violence. Thus, it is unlikely that gene therapy alone would completely eradicate depression; it would not rid us of the other factors, such as stressful life events, known to contribute to the problem of depression.

Even professional researchers have been known to put too much emphasis on single causes. For instance, the original version of the frustration-aggression theory of aggressive behavior stated that aggression is always caused by frustration (Dollard, Doob, Miller, Mowrer, & Sears, 1939). Today, psychologists know that this is not true; aggression has multiple causes. Factors such as pain, alcohol, or viewing media violence can also influence aggressive behavior (e.g., Berkowitz, 1983; Bushman & Cooper, 1990; Huesmann, 1998).

An overemphasis on single causes has also occurred, at times, in the longstanding nature-nurture debate over the causes of a variety of psychological phenomena. Introductory psychology textbooks (e.g., Myers, 2001) typically contain many examples of researchers who fell mainly on one side or the other of this debate. For example, Sir Francis Galton and Lewis Terman (developer of the Stanford-Binet Intelligence Scale) believed that intelligence was largely a matter of heredity rather than environment; behaviorists such as John B. Watson and B. F. Skinner believed that our behavior was mainly controlled by environmental forces; and researchers continue to debate the degree to which gender differences in our behavior are due to heredity or environment. However, many of our behaviors are influenced by *both* heredity and environment. Heredity and environment may not only produce main effects on our behaviors, but also may interact with each other to influence our behavior. The impact of the environment on our behaviors and reactions might depend upon our inherited predispositions. For example, a social gathering might cause those who have a genetic predisposition toward shyness to be more anxious than those without such a predisposition. Similarly, a virus-infested environment might be more likely to produce illnesses in those who have inherited a weak immune system compared with those with a strong immune system.

Nisbett and Ross (1980) explained that a tendency to focus on single causes is also related to the **overjustification effect,** which involves a reduction in a person's intrinsic (or internal) interest in an activity after being given extrinsic (or external) rewards for engaging in that activity. For instance, if you normally enjoy doing some activity (e.g., working on cars, cooking, or playing piano) and someone starts paying you money to do it, your intrinsic interest in the activity might drop sharply. Moreover, you might stop engaging in the activity when you are no longer getting paid. A number of psychological experiments have demonstrated this effect. For instance, Lepper, Greene, and Nisbett (1973) allowed nursery school children to play with "magic markers," an activity in which they had a lot of intrinsic interest. They told some of the children they would receive a nice award for playing with the colored markers; others were given no such external incentive. Two weeks later, the children were allowed to play in a room with a number of possible play activities, including the magic markers. Lepper et al. found that the children who were "bribed" with the extrinsic reward played with the markers only half as much as did the children who were given no extrinsic reward. Thus, the "bribed" children acted as if the extrinsic reward—rather than their intrinsic interest—caused them to play with the markers. Their intrinsic enjoyment of playing with markers may have been undermined by their focus on a single, external cause of their behavior; it was as if they assumed that if their behavior was affected by the external reward, it could not also be a result of their intrinsic interest.

Attribution Theory

Suppose that you saw a man named John slip and fall in a hallway of a shopping mall. If you also knew that (a) other people have walked down this hallway and none of them slipped, (b) John has slipped while walking down a number of other hallways, and (c) John has slipped in the same hallway a number of times in the past, what would you conclude about the cause of John's behavior? Why did he slip and fall in the hallway of the shopping mall?

Psychologist Harold Kelley (1967, 1973) proposed that people use an intuitive form of factorial ANOVA when they make attributions about the causes of events and behaviors. He said that people consider several classes of possible causes (or independent variables) when making such attributions: persons, entities (or situations), and times. How did you explain John's behavior? Did you conclude he slipped because he is clumsy (i.e., the person was the cause), the floor was slippery (i.e., the entity or situation was the cause), or because of something about the particular circumstances at that moment (i.e., the particular time was the cause)? Kelley said that we attribute behaviors (or effects) to the causes with which they covary (or co-occur). For example, John's behavior of slipping seems to occur across a variety of hallways and across time. Thus, the behavior of slipping covaries with John; this behavior "follows" him around wherever he goes. According to Kelley's attribution theory, you should conclude that John caused the slipping; perhaps he is a clumsy person. Referring back to the factorial ANOVA concepts discussed earlier, you might think of this as a main effect of the person (John).

Consider a different pattern of information related to our slipping example. Once again, you see John slip and fall in a hallway of a shopping mall. This time, however, you also know that (a) other people have walked down this hallway and all of them slipped, (b) you have seen John walk down other hallways and he never slipped, and (c) John has slipped in the same hallway in the past. What would you conclude about the cause of John's behavior? Why did he slip and fall in the hallway of the shopping mall? This time, the behavior of slipping covaries with the mall hallway. Everyone seems to slip in this hallway; there is a *consensus*. In this case, you should conclude that John's behavior of slipping in the mall hallway was caused by the hallway (i.e., the entity or situation); it appears to be a very slippery hallway. You might think of this as a main effect of the entity or situation (i.e., the hallway).

Consider a third pattern of information related to our slipping example. You witness John slipping and falling in the hallway, and you also know that (a) other people have walked down this hallway and none of them slipped, (b) you have seen John walk down other hallways and he never slipped, and (c) John has slipped in the same hallway a number of times in the past. Why did John slip and fall in the hallway of the shopping mall? Now it appears that John's behavior was caused by the unique combination of him and this particular hallway. You can think of this as an *interaction* between the person and the situation. In other words, the effect of the hallway (i.e., the situation) on a person's behavior *depends* on which person is walking down that hallway. When John is walking down the hallway, he slips and falls; when others walk down the hallway, they stay upright.

Biased Attributions

How did you answer Question 6 from the questionnaire? Did you think that you would help the student who was having a seizure? How did you view the person who failed to help? Did you think he was uncaring or cruel, even after you learned that his failure to help was a fairly common response among other people in the same situation? If so, you may be exhibiting a common attributional bias called the **correspondence bias** or **fundamental attribution error**. As discussed in Chapter 4, this bias is the tendency to draw inferences about a person's dispositions from behavior that can be completely explained by situational factors.

The study described in Question 6 was conducted by Darley and Latané (1968). Participants were led to believe that they were participating in a conversation with either one other student (i.e., the seizure victim), two other students (one of whom was the victim), or five other students. However, in actuality, each participant was alone; the other voices they heard over the intercom were tape recordings. Darley and Latané found that the size of the group had a dramatic impact on the number of participants who tried to help. The vast majority (85%) of those who thought they were alone with the victim tried to help before the intercom cut off his voice, but only 31% of those who believed there were five other students did so.[1] Thus, Darley and Latané found a main effect of the situation: Those who were in a situation with a larger group were much less likely to help than were those in a smaller group. Moreover, they reported that those who failed to help were not unconcerned; many of them were upset and asked about the victim when the experiment was over. Nevertheless, as we will see in a

---

[1]Sixty-two percent of the participants who believed there were five other students eventually reported the emergency after the intercom cut off the victim's voice. However, it took them a relatively long time to do so.

moment, people have difficulty acknowledging the effect of situational factors on the unhelpful behavior of Darley and Latané's participants (Nisbett & Borgida, 1975).

We have seen how people sometimes become too focused on single causes for events and behaviors. As mentioned in the discussion of attribution theory, Kelley (1967) hypothesized that we consider several potential causes for events and behaviors: persons, entities (or situations), and times. However, as implied by our discussion of the correspondence bias, we tend to focus mainly on only one of these causes: the person.

According to Kelley's theory, when you learn that many people behave in the same fashion in a situation (i.e., there is high consensus), this information suggests that the behavior is a result of situational forces. Thus, you should not conclude that their behavior is due to their unique personality. For example, when we know that most people slip and fall in a particular hallway, we should conclude that a person who slipped in that hallway did so because the floor was slippery, not because he is an unusually clumsy person. Similarly, if you know that failure to help a person in distress is a fairly common response when there are other bystanders around (e.g., Darley & Latané, 1968), you should assume that anyone who failed to help probably did so because of situational factors—that is, the presence of other bystanders—rather than personal factors. Nevertheless, Nisbett and Borgida (1975) found that this logic seems to escape people. They had college students read a description of the large-group condition (where participants conversed with five others) of Darley and Latané's study. Some of these students were given the actual results (i.e., percentage of people who helped) for this condition, and some were not. Of course, those who were given the results learned that the behavior of not helping was quite common (i.e., there was a high degree of consensus). Those who were not given these results assumed that this behavior was much less common. In one study, Nisbett and Borgida also gave the students a description of a man who never helped the seizure victim and then asked the students to (a) rate the extent to which his behavior was due to his personality or the situation and (b) rate him on several personality dimensions (e.g., apathetic/concerned, cruel/kind, and weak willed/strong minded). Although, logically, those who were given the consensus information—compared with those who were not—should have been less willing to make strong conclusions about the personality of the man who failed to help, they were not; the information that such behavior was typical made no difference in their ratings of the man.

The famous obedience experiments conducted by Stanley Milgram (1963, 1974) are another example of a surprising main effect of the situation on people's behavior. Milgram discovered that a majority of people will obey an experimenter's commands to deliver increasingly dangerous electric shocks to another person, despite the victim's screams, pleas to stop, and complaints of heart problems.[2] In fact, in his first study,

---

[2]For those unfamiliar with Milgram's studies, I should point out that the victim was an actor who was never truly shocked, and the participants were informed of this fact at the end of the experiment.

Milgram (1963) found that 65% of his participants delivered the maximum level of shock (450 volts) to the victim. Across a series of studies, Milgram convincingly demonstrated that changes in the situational features of the experiment—for example, the closeness of the experimenter to the participant, the status of the experimenter, and the presence of disobedient peers—affected the degree of obedience displayed by the participants. One might conceptualize this as a main effect of situational factors.

If you were told about a person who delivered the maximum level of shock in Milgram's (1963) experiment, what would you conclude about that person? Would you believe that person was aggressive or maladjusted? Because you know that a majority of Milgram's participants delivered the highest level of shock, I hope that you said you cannot attribute the person's behavior to his or her personality. However, the results of research by Miller, Gillen, Schenker, and Radlove (1974) suggest that people do attribute such behavior to the person's personality. They described Milgram's (1963) study to a group of college students. Half of these students were told that 65% of Milgram's participants administered the maximum level of shock, and half were not informed about this result. The students who were asked to report their impressions of a person who administered the highest level of shock—compared with students who rated a person who administered a low level of shock—gave relatively negative ratings of that person's personality, and the consensus information (i.e., Milgram's results) did not influence those ratings. Again, this is the correspondence bias at work: Behaviors that can be explained by situational factors are attributed to personal factors.

Decades of research have confirmed that the correspondence bias is a pervasive tendency (Gilbert & Malone, 1995). One possible consequence of this bias is that it might cause us to misperceive people who are largely victims of situational circumstances.[3] For example, we might believe that the Germans who participated in the slaughter of millions of Jews had to be especially cruel or insane. The results of Milgram's (1974) studies, however, suggest the possibility that we might have done the same thing in that situation. As another example, if we learn of several people who neglected to help a person experiencing a heart attack right in front of their eyes, we might perceive them to be particularly insensitive or cruel. Nevertheless, the research by John Darley and Bibb Latané (e.g., Latané & Darley, 1970) suggests that we might have behaved the same way in that situation. In these cases, it seems important that we recognize that causes other than personal factors influence the behavior of people; situational pressures also influence people's behaviors.

Summary

Factorial analysis of variance is a statistical procedure used by researchers to examine the effects of more than one independent variable on some outcome. This procedure allows a researcher to determine the separate main effects of each independent vari-

---

[3]See Gilbert and Malone (1995) for a more comprehensive discussion of potential consequences of the correspondence bias.

able as well as effects that result from the combination, or interaction, of more than one independent variable. The concepts of main effects and interactions are useful for reasoning about the causes of events and behaviors in our everyday lives. They may help us understand that events typically have multiple causes, even though we are sometimes tempted to focus only on a single cause. Psychologist Harold Kelley (1967, 1973) proposed that people use an intuitive form of factorial ANOVA when they make attributions about the causes of events and behaviors. He said that people consider several classes of possible causes when making such attributions: persons, entities (or situations), and times. However, researchers have found that we tend to focus on only one of those causes—persons—when we try to determine the cause of other people's behaviors. Even when another person's behavior can be explained by situational factors, people tend to conclude that the behavior reveals something about that person's unique inner qualities. This attributional bias, called the correspondence bias or fundamental attribution error, may lead to misperceptions of people whose behavior is largely due to situational pressures, such as those who obey orders to harm others or those who fail to help people during an emergency.

● PRACTICE PROBLEMS

1. We sometimes act as if we believe that only good people do good things and bad people do bad things. What does this imply about how we view the causes of others' behaviors? Explain how this is related to this chapter.

2. During a conversation with a group of your friends about various social problems, one of them says, "You know, if we would just love and respect one another, it would cure most of our social problems: People would not steal from each other, they would not kill each other, and they would be more willing to help each other." How would you respond?

3. Your next-door neighbor tells you that her son recently started college. He joined a fraternity and, during the initiation, he almost died. The members of the fraternity asked him to swallow a large slab of raw meat soaked in oil, and he nearly choked to death doing it. What impression would you form of your neighbor's son? Would your impression be different if your neighbor also told you that there was a large group of students being initiated into the fraternity that night, and all of them also agreed to swallow the meat?

4. A friend of yours tells you that she flunked a college course in psychological research methods, and this is the second time she flunked this course. You suspect that this outcome was caused by either a main effect of the person (i.e., your friend's academic ability is quite poor), a main effect of the situation (i.e., the course is very difficult), or an interaction between the person and the situation (i.e., there is something about the combination of your friend and this course that caused her poor performance). What information would you seek out to help you decide among these possibilities?

Analysis of Covariance

> Stereotypes might arise as a result of people's failure to engage in an intuitive analysis of covariance.
>
> *Schaller and O'Brien (1992)*

Imagine that you are a member of a golf league and your golf partner calls you the day before an important tournament to tell you that she is very sick. You need to find a new partner, and your current partner recommends two people—Diane and Jennifer—she knows well. She has played golf with both of these people numerous times in the past, and she gives you some information about the two players' performances. Before calling you, she glanced over their past scores and quickly categorized the scores as either "near par" (i.e., good) or "poor." She tells you that she has played the Hickory Links course with Diane eight times, and Diane scored near par on two of those occasions. She played this course with Jennifer twice, and Jennifer did not score near par either time. Moreover, she has played the Flatlands course with Diane two times, and Diane scored near par each time. Finally, she played the Flatlands course with Jennifer eight times, and Jennifer scored near par six of those times. Which player would you choose as your new partner? Why?

You might have noticed that for each golf course you were given a fairly small sample of golf rounds with which to judge the ability of Jennifer and Diane. After reading about the law of large numbers in Chapter 3, perhaps you decided to obtain a larger sample on which to base your judgment by aggregating the results across the two golf courses. If so, you would have noticed that Diane scored near par on a total of 4 out of 10 rounds (40%) and Jennifer scored near par on a total of 6 out of 10 rounds (60%). Would this lead you to pick Jennifer? Think about it for a moment, and we will come back to this example later.

This chapter will focus on **analysis of covariance.** We will start with a description of analysis of covariance and look at an example of how it is used by researchers. Then, we will discuss how the logic of analysis of covariance can be applied to everyday judgments, such as those we make about the personality traits of members of various groups.

What Is Analysis of Covariance?

Suppose that a researcher wants to compare the level of psychological stress in children raised primarily in single-parent homes with that of children raised primarily in two-parent homes. She locates fairly large samples of children from each type of home and has them complete measures of stress. As predicted, she finds that children raised in single-parent homes have higher levels of stress than do children raised in two-parent homes. Can she conclude that having only one parent causes higher levels of stress among children? As you might have guessed, other possible explanations exist for the different levels of stress in these two groups of children. Because the children were not randomly assigned to single-parent and two-parent homes, these two groups of children probably differ in ways other than the number of parents in their homes. For example, the single parents might have a lower socioeconomic status than the other parents. Thus, the children from single-parent homes might have higher levels of stress because of the lower socioeconomic status of their parents. From a researcher's point of view, the problem with this situation is that socioeconomic status is a **confounding variable.** In other words, the effect of the number of parents is confounded with—or cannot be distinguished from—the effect of socioeconomic status on stress. Fortunately, the researcher could use an analysis of covariance to handle this problem.

Analysis of covariance (ANCOVA) is a statistical procedure related to analysis of variance (see Chapter 7). Like analysis of variance, ANCOVA is used to evaluate differences among two or more groups of people. Unlike analysis of variance, however, ANCOVA allows researchers to statistically control for, or remove, unwanted variables—called covariates—that might be affecting the results of a study. For instance, the researcher mentioned above could use an ANCOVA to compare the stress levels of children raised by single parents with those of children raised by two parents while statistically removing the unwanted influence of socioeconomic status. If, after performing an ANCOVA, she still found that children from single-parent homes had higher levels of stress than did children from two-parent homes, she would be more confident that the difference was not due to socioeconomic status.[1]

---

[1]Although the researcher would be more confident that the different levels of stress in the two groups of children were not a result of socioeconomic status, she still could not conclude with certainty that the difference was a result of the number of parents in the home. Socioeconomic status might be one confounding variable that produced different levels of stress in the two groups of children, but other confounding variables may be present as well. For example, children in single-parent homes—compared with children in two-parent homes—may be more likely to have experienced parental divorce, and this might have increased their stress. Although ANCOVA can be used to control for more than one confounding variable, it is virtually impossible for an ANCOVA to control for all possible confounding variables (i.e., because the researcher might be unaware of them or unable to accurately measure them). Nevertheless, ANCOVA is very useful when the researcher can control for the confounding variables that seem most important to control.

Analysis of Covariance in Everyday Life

You might be wondering how this complex statistical procedure that magically removes confounding variables could be useful to you in everyday life. How could you possibly be expected to perform such complex calculations in your head? You may not be able to precisely calculate ANCOVA statistics in your head, but researchers (e.g., Schaller, 1992b; Schaller, Asp, Rosell, & Heim, 1996) have pointed out that an intuitive form of ANCOVA may be quite useful to us in our everyday lives. Simply put, an intuitive ANCOVA involves recognizing confounding variables and attempting to control for their influence whenever possible.

The following example will show how you might use an intuitive form of ANCOVA. Table 8.1 presents a summary of the data from the golf example at the beginning of the chapter. (This example is a modified version of one developed by Schaller, 1992b.) As I mentioned earlier, when you look at the golfers' totals across the two golf courses, you find that Diane scored near par on a total of 4 out of 10 rounds (40%) and Jennifer scored near par on a total of 6 out of 10 rounds (60%). These figures might lead you to conclude that Jennifer is a better golfer than is Diane. However, a close look at Table 8.1 will reveal that Diane played most (8 out of 10) of her rounds on the Hickory Links course, and Jennifer played most (8 out of 10) of her rounds on the Flatlands course. In other words, golfer and golf course are confounded with one another. This would not be a problem if the two golf courses were equally difficult, but the fact that both golfers performed much better on the Flatlands course suggests that Hickory Links is the more difficult course.

In order to get a more accurate sense of whether Jennifer or Diane is the better golfer, we would have to control for the effect of the golf course because the course imposes a situational constraint on the players' performances. One way to do this is to simply hold the golf course constant while you compare the two golfers' performances.

Table 8.1

Hypothetical Golfers' Performances at Two Golf Courses

Golfer	Hickory Links		Flatlands		Total	
	Near Par	Poor	Near Par	Poor	Near Par	Poor
Diane	2 (25%)	6	2 (100%)	0	4 (40%)	6
Jennifer	0 (0%)	2	6 (75%)	2	6 (60%)	4
Total	2 (20%)	8	8 (80%)	2		

Note: A player's near-par percentage, shown in parentheses, is the number of near-par rounds divided by the total number of rounds for a course (e.g., Diane's near-par percentage for the Hickory Links course is 2/8, or 25%).

In other words, you could compare the performances of Diane and Jennifer on the Hickory Links course, and then compare their performances on the Flatlands course. The first comparison reveals that Diane scored near par on 25% of her rounds at Hickory Links, and Jennifer never scored near par on this course; therefore, Diane had a better performance on this course. When you compare the two golfers' performances on the Flatlands course, you see that Diane scored near par on 100% of her rounds and Jennifer scored near par on 75% of her rounds; once again, Diane outperformed Jennifer on this course. Once we control for the confounding variable (the golf course), it becomes clear that Diane is the best person to choose as your new golf partner. The only reason that Jennifer's overall performance appeared better than Diane's is that Jennifer played most of her rounds on an easier golf course. When we compare the two golfers' performances on the same course, Diane outperforms Jennifer each time.

If you failed to perform an intuitive ANCOVA when you first read this problem and you concluded that Jennifer was the better golfer, you are not alone. Schaller (1992b) gave college students a similar problem, but one that involved racquetball performance instead of golf performance. He found that most of them concluded that the player with the better overall performance was the best player, and this result suggests that they did not use an intuitive ANCOVA. However, when the students were given data containing much larger samples of the players' performances (i.e., 10 times the number of games presented in the original version), most of them picked the other player; thus, their reasoning was consistent with an intuitive ANCOVA. Schaller suggested that the students who received the smaller sample sizes felt the need to aggregate the data—because larger samples give better estimates than smaller samples—and base their judgments on the players' overall performances. Ironically, it appears that the students' understanding of one statistical concept (the law of large numbers) interfered with their ability to apply a second statistical concept (ANCOVA).

ANCOVA and Stereotyping

Now that you know a little about ANCOVA, reconsider Problem 7, originally presented in Chapter 1:

7. Imagine that we tested the verbal intelligence of members of two different racial groups (Group A and Group B). Both groups were given 25 anagrams to solve, some of which were five-letter anagrams and some seven-letter anagrams. Members of Group A solved all 5 (100%) of the five-letter anagrams and 5 (25%) of the 20 seven-letter anagrams they were given (see Table 8.2). Members of Group B solved 15 (75%) of the 20 five-letter anagrams and none (0%) of the 5 seven-letter anagrams they were given. Considering their performance on the anagrams, which group would you say has the highest verbal intelligence?

Table 8.2

Anagram Performance of Two Groups

	Five Letters		Seven Letters		Total	
Group	Solved	Unsolved	Solved	Unsolved	Solved	Unsolved
Group A	5 (100%)	0	5 (25%)	15	10 (40%)	15
Group B	15 (75%)	5	0 (0%)	5	15 (60%)	10
Total	20 (80%)	5	5 (20%)	20		

Note: The percentages shown in parentheses are the number of solved anagrams divided by the total number of anagrams of a particular type (e.g., Group B solved 75% of the 20 five-letter anagrams). Adapted with permission from "'Intuitive Analysis of Covariance' and Group Stereotype Formation," by M. Schaller and M. O'Brien, 1992, *Personality and Social Psychology Bulletin, 18,* p. 778. Copyright 1992 by the Society for Personality and Social Psychology, Inc.

Schaller and O'Brien (1992) showed college students the anagram performance information contained in Question 7. However, they presented the results for all 50 anagrams individually, one anagram at a time (showing the anagram and which group solved it). They found that the students judged Group B to be more intelligent than Group A. Did you draw the same conclusion? If you look at the overall performance of both groups, Group B solved 15 of their anagrams and Group A solved only 10 of their anagrams. These data suggest that Group B is more intelligent than Group A. However, as you might have guessed, this conclusion is inaccurate because it does not control for the different types of anagrams attempted by these groups.

Both groups performed more poorly on the seven-letter anagrams than on the five-letter anagrams, which suggests that the seven-letter anagrams were more difficult. Moreover, Group A attempted mainly seven-letter anagrams and Group B attempted mainly five-letter anagrams. Therefore, the superior overall performance of Group B may be due to the fact that they had easier anagrams. To control for this confounding variable and perform an intuitive ANCOVA, you can compare the performances of the two groups while holding the type of anagram constant. If you look at the groups' performances on the five-letter anagrams, Group A outperformed Group B (i.e., 100% versus 75%). On the seven-letter anagrams, Group A outperformed Group B once again (25% versus 0%). Thus, once you take into account the contaminating effects of anagram type, it becomes clear that the members of Group A exhibited higher verbal intelligence than did the members of Group B. Because Schaller and O'Brien's (1992) participants failed to perform an intuitive ANCOVA and control for the situational constraints (i.e., anagram difficulty) on group members' performances, they inaccurately "stereotyped" Group A members as less intelligent.

Although the anagram example is fictitious, some real-world examples of stereotypes might be a result of our failure to control for situational constraints on group members' behaviors. For example, Schaller and O'Brien (1992) pointed out that we might believe that African Americans are less intelligent than whites because African Americans typically score lower than whites on standardized tests such as achievement tests. They explained, however, that the correlation between race and test performance "may be entirely misleading, resulting from the fact that socioeconomic status (among other variables) is powerfully confounded with both race and achievement test performance" (p. 777). In other words, the different test performances of African Americans and whites might be due to the difference in socioeconomic status between these two groups, not the difference in race.

Stanovich (2001) noted that socioeconomic status is one of the important confounding variables that might account for the higher achievement of students in private, compared with public, schools. The relationship between type of school and achievement may lead us to infer that private schools offer a higher quality education than do public schools. The problem with this inference, however, is that it ignores the fact that the children who attend public schools tend to have a lower socioeconomic status than do children who attend private schools. Once you control for this and other confounding variables, you find similar levels of student achievement in private and public schools (e.g., Page & Keith, 1981).

Our failure to take into account situational influences on the behavior of different groups of people might also lead us to develop gender stereotypes. For example, women are typically viewed as more submissive and compliant than are men. Eagly and Steffen (1984) and Schaller (1992a) suggested that this stereotype might be a result of our failure to take into account the fact that men and women typically hold very different social positions. Men are more likely than women to hold social positions that are high in status and authority; women are more likely than men to hold low-status positions. High-status positions are more likely to require assertive, directive behaviors; low-status positions are more likely to require submissive, compliant behaviors. Thus, if we were to count up the overall number of assertive behaviors we have observed in men and women, we would probably find a greater number among men. But this may be due to the fact that men are more likely than women to have high-status social positions. We would have to compare the behavior of men and women who hold *similar* social positions in order to control for this confounding variable.

Are people more likely to perform an intuitive ANCOVA, thereby avoiding negative stereotypes, when the negative stereotypes involve their own group? For example, are women more likely than men to take into account the effects of social position on people's behavior? Schaller (1992a) investigated this question by asking men and women to determine the relationship between gender and leadership ability from a list of 40 employees (20 males and 20 females) who ostensibly worked for an aerospace company. For each employee, this list indicated the person's position—executive or office worker—in the company and his or her leadership ability (i.e., good or bad). The list was constructed in such a way that (a) executives were more likely to be good leaders than were office workers, (b) men were more likely to be executives than were women, and (c) men were more likely to be good leaders than were women. The relationship between gender and leadership ability, however, was due solely to the fact that men were more likely to be executives and executives were more likely to be good leaders. Notice how these hypothetical data parallel the real-life confounding of social position and gender discussed above. Schaller found that the female participants in his study were more likely than the male participants to conclude that the list showed no relationship between gender and leadership ability. The majority of the male participants concluded that the information showed that men were better leaders than were women; however, the majority of the female participants concluded there was no gender difference in leadership ability. Therefore, it appears that the female participants were more motivated to use an intuitive ANCOVA—thereby avoiding the negative stereotype that women are poor leaders—than were the male participants.

If the failure to use an intuitive ANCOVA can lead to the development of inaccurate stereotypes, perhaps training people to use this form of statistical reasoning would prevent such stereotypes. Schaller, Asp, Rosell, and Heim (1996) trained a group of college students to use the logic of ANCOVA to control for confounding variables. About a week later, these students participated in a study in which they were asked to judge the intelligence of two groups of people based on their performance on anagrams (the same task, described on p. 107, used by Schaller & O'Brien, 1992). Remember that Schaller and O'Brien found that people who were given this task judged Group B—the group with the best overall performance, but with the easiest anagrams—to be more intelligent than Group A. Schaller et al., however, found that the students who had received ANCOVA training made more accurate judgments and rated Group A as more intelligent than Group B. A second, untrained group of participants in Schaller et al.'s study failed to take into account the contaminating effects of anagram difficulty; they, like the participants in Schaller and O'Brien's study, judged Group B as more intelligent than Group A. Thus, training in the logic of ANCOVA appears to make people more likely to recognize confounding variables and control for them when making judgments, and this ability might make people less likely to form erroneous group stereotypes. Perhaps you will experience similar benefits from reading this chapter.

ANCOVA and Perceptions of Changes in Subpopulations

Suppose that you live in a fairly large city and, one morning while reading your local newspaper, you see a story on the problem of teenage pregnancy in your area. The author of the article reports that a nearby high school had 1,200 students in 1990, and in that year there were about 25 pregnancies among the teenage women at that school. By the year 2000, the number of teenage pregnancies had increased to 50. The author states, "As you can see, the rate of teenage pregnancies at our local high school has doubled in the past 10 years. I do not know what has caused this problem—maybe this is yet another result of the moral decay in our society—but I do know that this problem is much more serious than it used to be, and it is time we did something about it." Would you agree with the author's conclusion that the problem is "much more serious than it used to be"? If you could ask for additional information before answering this question, what would you ask for? One important piece of information you might want is the size of the student population in 2000. If you knew that the student population increased from 1,200 in the year 1990 to 2,400 in the year 2000, would you conclude that the pregnancy problem had become much more serious? Using these population figures, you can see that the *rate* of pregnancies in 1990 was about 2% (25/1,200), and it remained at 2% (50/2,400) in the year 2000. Thus, although the *number* of pregnancies doubled between 1990 and 2000, the pregnancy *rate* (or percentage) remained the same. Thus, the problem was not any more serious in 2000 than it was in 1990. The size of the student population was a confounding variable that the author neglected to consider before drawing conclusions about the pregnancy problem.

In the pregnancy example, controlling for the size of the population—in the spirit of an intuitive ANCOVA—is a simple matter of dividing the number of pregnancies by the size of the student population to determine the percentage of students who were pregnant in each year. However, Silka and Albright (1983) found that people have difficulty taking into account population information unless that information is made very conspicuous to them. In one study, they asked students to evaluate the effects of a sex education program at a high school and presented them with information about changes in the number of pregnancies at the school over a 10-year period. Some of the participants were told that the number of pregnancies had doubled during this time period; others were told it had stayed the same. Participants also received, embedded within a description of the school, information about the size of the student population during the same period. Some of the participants were informed that the student population decreased by 50%, others were told it stayed the same, and a third group was told that the student population had doubled. The participants who learned that the *number* of pregnancies doubled during that time period indicated that the pregnancy *rate* had increased substantially, regardless of the population sizes they had been given. If the participants had taken into account the population size, those who learned that the population size doubled along with the number of pregnancies should have known that the pregnancy rate remained unchanged.

In another study, Silka and Albright (1983) omitted the population information—as might happen in media accounts of social problems such as teenage pregnancy—and gave the participants an opportunity to request additional information to assess the seriousness of the pregnancy problem, but very few of the participants requested information about the population sizes. It was not until Silka and Albright put a greater emphasis on the information about population size (in a third study) that their participants' judgments were influenced by that information.

The study by Silka and Albright (1983) reminds me of an event that takes place every year at my college. One week before the start of the school year, the faculty and staff gather to watch a presentation by the director of admissions. Before the 1999–2000 school year, he informed us that the class of incoming students was bigger than it had been the previous year. In fact, this was the biggest group of incoming students in the history of the college. He congratulated the admissions staff who had worked so hard out there in the "trenches" of high schools and college fairs to recruit students. He complimented the faculty on the fine job they performed, resulting in satisfied students who spread the word about the great education they received at our college. We all felt elated and proud that our hard work had paid off; more students were choosing our college.

This is a familiar scene at my college each year, at least during the years that we experience increased enrollments. But I have often wondered if we were all taking a bit too much of the credit for the increases in enrollment. Might there be some other explanation for such increases—perhaps a confounding variable that we were overlooking? Several factors probably affect enrollments at colleges, but one obvious factor is the total number of high school graduates who plan to go on to college. If the population of potential college students increased from one year to the next, we would probably experience an increase in our enrollment. In fact, the number of high school graduates in Ohio did increase from 1998 to 1999; this increase appears to account for most of the increase in my college's enrollment during the same time period.

Perhaps we should have thanked the forces that affect the total number of potential college students (e.g., birth rates, high school graduation rates, and economic forces) rather than patting ourselves on the back. An intuitive ANCOVA that controls for the contaminating effects of increases in the population of potential college students can be easily carried out in this situation; we could simply divide the number of incoming freshmen by the total number of potential college students in our area to arrive at the percentage of available students who chose our college. If, during a particular year, we had a greater *number* of incoming freshmen, but not a greater *percentage* of the available students (i.e., the increased number of freshmen was due to increases in the total population of students), I am sure that we would still be happy—but we would be less tempted to infer that we all deserve a pat on the back.

Summary

Analysis of covariance (ANCOVA) is a statistical procedure that researchers sometimes use to control for confounding variables when assessing relationships between events or variables. We can use an intuitive form of ANCOVA by recognizing and taking into account confounding variables that might distort our judgment of the relationships we observe between events in our everyday lives. For example, if we observe men acting differently than women, we might consider the extent to which this is a result of the differing situational constraints or social positions in which the two sexes often find themselves. Comparing the behaviors of men and women in the same situation would help control for the contaminating effects of situational constraints. An intuitive ANCOVA might also help us make more accurate judgments about changes in subpopulations if we control for changes in the size of the entire population of interest. For instance, if we hear that pregnant teenagers are becoming more numerous, we can use information about changes in size of the teen population as a whole to determine if there has been a change in the rate as well as the number of pregnancies. Training in the logic of ANCOVA appears to make people more likely to recognize confounding variables and control for them when making judgments. One potential benefit of such training is that those who receive it might be less likely to form erroneous group stereotypes.

PRACTICE PROBLEMS

1. Two friends of yours, Kristy and Jenny, are arguing over who is the better college student, and they have asked you to give your opinion. They have both taken psychology courses and communications courses, but Kristy is a psychology major and Jenny is a communications major. Jenny points out that she received As in 5 out of the 8 psychology and communications courses she has taken, and she received Bs in the remaining courses. Jenny also tells you that Kristy received As in only 3 out of the 8 psychology and communications courses she took, and she received Bs in the rest. Kristy points out that she took 2 communications courses and received As in both of them, and she took 6 psychology courses and received an A in 1 of them. Moreover, Kristy mentions that Jenny took 6 communications courses and received As in 5 of them; Jenny also took 2 psychology courses and received Bs in both of them. Who would you say is the better student? Why?

2. A recent newspaper article implied that college professors are making too much money. To prove his point, the author compared the average salary for full professors in Michigan to the average salary for all workers in Michigan. The average salary for full professors was much higher than that of all Michigan workers. Explain the problem with the author's analysis and describe what you would do to improve it.

3. While listening to a news report, you learn that working women are earning less money, on the average, than are working men. Before you conclude that the difference in pay is a result of discrimination against women, what additional information do you need?

4. You read a newspaper article in which the author states that Chicago is a much more dangerous place than is Detroit. She drew this conclusion after learning that the number of murders within the city limits of Chicago was much higher than the number of murders that occurred within the city limits of Detroit. Would you agree with the author's conclusion? Explain your answer.

Conditional Probability and Bayes' Theorem

> The discrepancy between what clients believe a positive HIV test means and what it actually does mean seems to have cost human lives in addition to the toll the disease itself has taken.
>
> *Gigerenzer, Hoffrage, and Ebert (1998)*

Let's revisit a scenario presented in Chapter 1: After donating blood, you learn that you tested positive for the AIDS virus (HIV). Feeling completely shocked and devastated, you quickly tell your partner you have to leave, and you drive to the blood center to speak with a physician. "There must be some mistake," you tell the physician. "I feel completely healthy, I have not had sex with anyone who has AIDS, and I have never used intravenous drugs." The physician responds, "The test is extremely accurate; if a person has HIV, the test detects it in 99.8% of the cases." "So," she continues, "I'm sorry, but the probability is practically 100% that you have HIV." You drive home in tears, wondering what the future holds. Will your romantic partner, whom you were about to marry, leave you? Will you lose your job? How long will you live?

After some careful reading on the topic of AIDS, you discover that the rate of HIV infection in people like yourself—individuals who do not engage in risky behaviors, such as intravenous drug use—is about 1 in 10,000 (this is referred to as the **prior probability,** or base rate, of the disease). The low base rate of the disease suggests that it is highly improbable that a person like you would have HIV. However, you also learn that the physician was correct about the "accuracy" of the test; among individuals who have HIV, the test detects it in 99.8% of the cases (this is called the **sensitivity,** or hit rate, of the test). Finally, you discover that the test incorrectly produces a positive result in only .01% of individuals who do not have HIV (this is referred to as the **false positive rate**).[1]

As shown in Chapter 1, the physician was incorrect in concluding that the "probability is practically 100% that you have HIV." The physician should have told you

---

[1] The figures I presented for the base rate, sensitivity, and false positive rate are the same as those used by Gigerenzer, Hoffrage, and Ebert (1998). These figures are based on research studies that have examined actual HIV rates and test results.

that the probability you have HIV is approximately 50% (i.e., this is the probability that you have HIV, given that you tested positive, or the **posterior probability**). So, it is certainly possible that you do *not* have HIV; in other words, it is possible that the test result was a false positive. Although the situation I asked you to imagine is fictional, it is far from impossible. Researchers have discovered that people, including physicians, make this type of mistake in statistical reasoning when interpreting the results of laboratory tests (e.g., Casscells, Shoenberger, & Graboys, 1978; Gigerenzer & Hoffrage, 1995; Gigerenzer, Hoffrage, & Ebert, 1998; Hoffrage & Gigerenzer, 1998). Mistakes of this sort could have tragic results. For example, individuals from low-risk groups who test positive for HIV after donating blood might commit suicide because they are convinced it is an absolute certainty that they have the virus. In fact, Gigerenzer et al. (1998) described an example of 22 blood donors in Florida who tested positive for HIV, 7 of whom later committed suicide. (See the quote that opened this chapter.)

Are you surprised that the posterior probability in the HIV example is 50% instead of 100%? This chapter will examine why the answer is 50%. First, the topics of conditional probability and Bayes' theorem will be discussed. The chapter will also look at how these statistical concepts are relevant to our everyday lives, including their relevance to decisions we make in medical and legal settings. Finally, some common difficulties people have in Bayesian reasoning will be examined.

Understanding Conditional Probability and Bayes' Theorem

In the HIV example, a proper interpretation of your positive test result requires that you determine if the positive result means that you actually have HIV. In other words, you have to determine the probability that you actually have HIV *given* the positive test result. This is called a **conditional probability**—the probability of some event, A, given that another event, B, has occurred. Researchers often use the notation $p(A \mid B)$ to represent the probability of A given B (i.e., "*p*" stands for "probability" and the "|" means "given"). For the HIV example, you need to determine $p(\text{HIV} \mid \text{positive result})$.

Although the blood test indicated that you have HIV, you would probably not instantly conclude that you actually have the virus. You would probably be shocked by the test result because you know that HIV is extremely rare among low-risk individuals; earlier, we saw that the base rate of the disease in this group was about 1 in 10,000. If you were an intravenous drug user who had shared needles with people known to have AIDS, you might be more willing to believe the positive result of the HIV test. Because you are in a low-risk group, however, one of your first reactions might be to question whether the test result was a mistake (i.e., a false positive). Obviously, if you learned that the test always indicates a positive result among people who do not have HIV (i.e., false positive rate = 100%), you would not believe your test result. Likewise, if you learned that the test indicates a positive result among only 40% of people who do have HIV (i.e., sensitivity = 40%), you might not trust the test. Thus, base rates, false positive rates, and sensitivity are all important factors to consider in determining the probability that you have HIV given a positive test result. Researchers rely on **Bayes' theorem** to determine how to combine all of this information to determine conditional probabilities. For the sake of completeness, the footnote below demonstrates the use of a Bayesian equation to solve the HIV problem.[2] However, for our purposes, a more intuitive approach to this problem will work just fine. We will translate the base rate, sensitivity, and false positive rate into numbers of people, or what Gerd Gigerenzer and his colleagues call "natural frequencies" (e.g., Gigerenzer & Hoffrage, 1995).

As mentioned above, the base rate of HIV among low-risk individuals is about 1 in 10,000. Thus, if we tested 10,000 people for HIV, 1 person would actually have the virus. Because the sensitivity of the test is 99.8%, it is a virtual certainty that this HIV-infected person would test positive for HIV. Moreover, of the 10,000 people tested, 9,999 would not have HIV. But, because the false positive rate of the test is .01%, we would expect 1 of these 9,999 healthy people (i.e., .01% × 9,999 = 1) to test positive for HIV. Therefore, of the 2 individuals who tested positive for HIV, only 1—or 50%—actually has the virus (see Table 9.1). In other words, the probability that a low-risk individual actually has HIV given a positive test result (i.e., the posterior probability) is 1/2, or 50%. In general, the posterior probability can be easily determined if you divide the number of true positives (in this case, people who actually have HIV and tested positive) by the total number of true positives and false positives [i.e., posterior probability = true positives/(true positives + false positives) = 1/(1 + 1) = 1/2 = 50%].

For members of high-risk groups, the probability that a positive test result means that they actually have HIV is much higher. For instance, assume you are a member of a group in which 1 out of every 100 individuals is infected with HIV (i.e., base rate = 1%) and your HIV test result is positive. What is the probability that you actually have HIV given this positive test result [i.e., $p(\text{HIV} \mid \text{positive result})$]? If 10,000 of these individuals were tested, 100 (or 1%) of them would actually have HIV. Because the

---

[2]One version of Bayes' theorem is as follows:

$$p(\text{HIV} \mid \text{pos}) = \frac{p(\text{HIV})p(\text{pos} \mid \text{HIV})}{p(\text{HIV})p(\text{pos} \mid \text{HIV}) + p(\text{no HIV})p(\text{pos} \mid \text{no HIV})}$$

where $p(\text{HIV} \mid \text{pos})$ = probability you have HIV given a positive test result,
$p(\text{HIV})$ = base rate of HIV (1 in 10, 000, or .0001),
$p(\text{pos} \mid \text{HIV})$ = sensitivity or the probability of a positive test result given the presence of HIV (99.8%, or .998),
$p(\text{no HIV})$ = $1 - p(\text{HIV})$ (9999 in 10,000, or .9999), and
$p(\text{pos} \mid \text{no HIV})$ = false positive rate (.01%, or .0001), or the probability that a person will test positive given that he or she does not have HIV.

$$p(\text{HIV} \mid \text{pos}) = \frac{.0001(.998)}{.0001(.998) + .9999(.0001)} = \frac{.0000998}{.0000998 + .00009999} = .5, \text{ or } 50\%$$

Table 9.1

Frequencies Reflecting Rates of HIV Infection and HIV Test Results

	Test Results		
Actual HIV Status	Positive	Negative	Total
HIV Present	1	0	1
HIV Absent	1	9,998	9,999
Total	2	9,998	10,000

sensitivity of the test is 99.8%, virtually all of these 100 people would test positive for HIV. Thus, the number of true positives is 100. Of the remaining 9,900 people who do not have HIV, .01% or 1 of them would test positive for HIV; in other words, the number of false positives would be 1. If we divide the number of true positives by the total number of true and false positives, we find that the probability is 99% that you actually have HIV given the positive test result [i.e., p(HIV | positive result) = true positives/(true positives + false positives) = 100/(100 + 1) = 100/101 = 99%). As you can see, when the base rate of a disease is higher, a positive test result is more likely to be a true positive.

Common Difficulties in Bayesian Reasoning

If you are like most people, you probably had trouble determining the probability that you actually had HIV given a positive test result when you first encountered the HIV example. A number of researchers (e.g., Casscells, Schoenberger, & Graboys, 1978; Tversky & Kahneman, 1982a; Gigerenzer et al., 1998) have found that people have difficulty with **Bayesian reasoning.** For example, Gigerenzer et al. had a member of their research team pose as a low-risk, heterosexual client, and he went to 20 public AIDS counseling centers in Germany. They wanted to see if the centers' counselors—70% of whom were physicians and 30% social workers—would accurately explain what a positive HIV test result means. They found that all of the counselors vastly overestimated the probability that a low-risk person has HIV given a positive test result. Most of them told the client that this probability was 100%, and the rest of the counselors indicated it was 90–99.9%. As shown earlier, the probability is actually 50%. Why would these professionals be so far off in their probability estimates? We are going to take a look at two of the reasons why people seem to have difficulty with Bayesian reasoning: confusion about conditional probabilities and base-rate neglect.

Confusion About Conditional Probabilities

One of the errors exhibited by the counselors in Gigerenzer et al.'s (1998) study was the assumption that the probability of HIV given a positive test result [p(HIV | positive result)] was the same as the probability of a positive test result given HIV [p(positive result | HIV)]. The probability of a positive test result given the presence of HIV is the sensitivity of the test. As mentioned earlier, if a person has HIV, the test detects it in about 99.8% of the cases. Thus, p(positive result | HIV) = 99.8%. A number of the counselors suggested that this value was also the probability of having HIV given a positive test result [i.e., p(HIV | positive result) > 99%]. But, as we have seen, this probability is actually 50%.

Although it is a common mistake to assume that a conditional probability is the same as its inverse, the two are not identical. Obviously, the probability that an animal has four legs given that it is a cow is not the same as the probability that an animal is a cow given that it has four legs (i.e., all sorts of animals have four legs). Likewise, the probability that a woman lives with someone given that she is married is not the same as the probability that she is married given that she lives with someone (i.e., that person could be a friend or college roommate). These examples are clear-cut, but conditional probabilities are easier to confuse in other situations. For instance, Dawes (1988) described a newspaper story that stated that most drug users also use marijuana, but the headline read, "Most on Marijuana Using Other Drugs" (p. 71). The headline referred to the probability that a person uses other drugs given that the person uses marijuana [p(other drugs | marijuana)], but the story itself was about the probability that people use marijuana given that they use other drugs [p(marijuana | other drugs)]. The percentage of marijuana users who use harder drugs is probably much smaller than the percentage of hard-drug users who use marijuana (see Dawes). Another example was presented as Question 8 in the questionnaire in Chapter 1:

> **8. Suppose that you learned in your introductory psychology class that most of the people who have committed suicide talked about doing it beforehand. Would you infer from this information that a person who talks about suicide is very likely to commit suicide?**

If so, you are confusing the two conditional probabilities. You learned in your introductory psychology class that there is a high probability that a person talked about suicide given that he or she committed suicide (e.g., Meyers, 2001). But the probability of suicide given that a person talked about it is much lower; few of these people actually commit suicide (Yip, 1998). Note that this does not imply that should you ignore a person who tells you that he or she is thinking of committing suicide. Although there may be a low probability that this person will actually commit suicide, he or she may need professional help.

Among those who interpret the results of medical tests, confusing conditional probabilities could result in a serious mistake. As shown above, researchers have found that physicians sometimes confuse the probability of the disease given the test result with the probability of the test result given the disease (e.g., Eddy, 1982; Gigerenzer et al., 1998). Eddy found that physicians assumed that the probability of breast cancer given a positive mammogram (i.e., X-ray) was about 75%, but a proper Bayesian analysis of the information they were given showed it was actually 8%. This dramatic error in statistical reasoning could have disastrous results if a woman who is cancer-free is misdiagnosed as having breast cancer.

Base-Rate Neglect

Imagine that you are a juror listening to a case about a cab involved in a hit-and-run accident. During the case, you learn that only two cab companies—Green and Blue—operate in the city. You also learn the following:

a. Eighty-five percent of the cabs in the city are Green, and fifteen percent are Blue.
b. A witness identified the cab as Blue. The court tested the reliability of the witness under the same circumstances that existed on the night of the accident and concluded that the witness correctly identified each one of the two colors 80% of the time and failed 20% of the time.

What is the probability that the cab involved in the accident was Blue rather than Green? Tversky and Kahneman (1982a, pp. 156–157), who originally developed this cab example, found that their participants estimated the probability to be around 80%. It is actually 41%. To understand why it is 41%, apply the logic of Bayesian reasoning discussed earlier. We want to know the probability that the cab is Blue given the identification by the witness [i.e., p(Blue | witness identification)]. Suppose there are 100 cabs in this city, and 15 are Blue cabs. Because the witness is 80% reliable, we would expect him or her to correctly identify 12 of these 15 cabs as Blue (this is the number of true positives). Of the 85 cabs in the city that are Green, the witness would incorrectly identify 20%—or 17—of them as Blue (this is the number of false positives). Therefore, of the 29 cabs (i.e., 12 + 17 = 29) identified as Blue by the witness, 12—or 41%—would actually be Blue [i.e., p(Blue | witness identification) = 41%].

Tversky and Kahneman (1982a) suspected that their participants overestimated the probability that the cab was Blue because they did not adequately consider the fact that only 15% of the cabs in the city were Blue. In other words, they suspected that people were underutilizing this base-rate, or population, information in favor of the case-specific information concerning the witness's identification of the cab as Blue. They called this tendency **base-rate neglect**. After all, who cares that there is a small percentage of Blue cabs in the city when a witness who saw the accident identified the cab as Blue, right? Wrong. As shown in the HIV problem, the base rate can have a big impact on the probability that a positive result, or positive identification, is a true positive. The relevance of the base rate to your judgment in the Blue cab example becomes obvious when you consider the extreme base rates of 100% and 0% Blue cabs.

Obviously, you would be much more likely to be influenced by the witness's statement that the cab was Blue if you learned that 100% of the cabs in the city were Blue than if you learned that 0% were Blue.

Tversky and Kahneman (1982a) found that people were more likely to utilize the base-rate information in the cab problem when the base rate was modified to make it seem more relevant to the hit-and-run accident. Specifically, rather than simply mentioning the percentage of Green and Blue cabs in the city, they told participants, "Although the two companies are roughly equal in size, 85% of cab *accidents* [emphasis added] in the city involve Green cabs and 15% involve Blue cabs" (p. 157). This information seems very relevant to an assessment of the likelihood that a Blue cab caused the accident. The participants who received this information gave an average probability estimate of 60%. Although this estimate is still not equal to the Bayesian-derived estimate of 41%, it is much lower than the 80% figure given by most of the participants who received the original base-rate information. Other researchers have also found that making base rates seem more relevant to individual cases increases the chance that people's judgments will be influenced by base-rate information. For example, Brekke and Borgida (1988) had college students serve as mock jurors in a rape trial, and some of them heard an expert discuss rape statistics (i.e., base rates, such as "most women are reluctant to report rapes to the authorities," p. 385). In one condition, the expert pointed out the connection between the rape statistics and a specific rape case (e.g., "The fact that she had to be talked into reporting the incident is not at all unusual or hard to understand," p. 385); in another condition he discussed only the statistics. The researchers found that the rape statistics had a larger impact on jurors' judgments of the case when the expert explained how they were relevant to a specific case.

Base-rate neglect has also been found in the realm of medical testing. As part of the study by Gigerenzer et al. (1998) in which a member of their research team visited AIDS counseling centers in Germany, they asked counselors about the prior probability, or base rate, of HIV in heterosexual men with low-risk behavior. Most of the counselors were either uncertain or ignorant of the base rate. Some of them even claimed that base rates were of little use in determining the meaning of an individual's HIV test result; one said, "Statistics don't help us in the individual case . . . " (p. 204). Clearly, this statement illustrates a blatant form of base-rate neglect. Similarly, Eddy (1982) suggested that physicians are confused about the relevance of base rates to clinical diagnoses of individuals. For instance, he quoted a textbook on clinical diagnosis as saying, "Statistical methods can only be applied to a population of thousands. The individual either has a rare disease or doesn't have it; the relative incidence of two diseases is completely irrelevant to the problem of making his diagnosis" (DeGowin & DeGowin, 1969, p. 6). It is true that an individual either has a rare disease or doesn't have it, but the incidence of the disease is completely relevant to the task of determining the *probability* that he or she actually has the disease given a positive diagnostic test result. In the HIV examples given earlier, you saw that when the HIV base rate was 1 in 10,000, the probability that a person had HIV given a positive test result was 50%; this probability increased to 99% when the base rate was 1 in 100. The difference between 50% and 99% would be very relevant to a person faced with a positive HIV test result.

Previous chapters discussed additional examples of people's tendency to underutilize base rates, or population information, in favor of case-specific information. Chapters 3 and 4 showed that people sometimes base their judgments on single-case evidence, even when they have group statistics (i.e., base rates). For example, in the study by Hamill, Wilson, and Nisbett (1980), the participants who read a negative description of one welfare recipient reported more unfavorable attitudes toward welfare recipients in general than did the participants who did not read the description; and this occurred regardless of whether they were told that the woman's lengthy stay on welfare was typical or atypical. The information about typicality was base-rate information, and the participants largely ignored it. Chapter 7 discussed a study by Nisbett and Borgida (1975) in which the participants' judgments of a person who failed to help a seizure victim were just as harsh when they were told this behavior was common (i.e., a high base rate) as when they assumed it was rare. Thus, people's tendency to underutilize base-rate information is a factor in several of the difficulties in statistical reasoning considered in this book.

Improving Our Ability to Engage in Bayesian Reasoning

Engaging in Bayesian reasoning seems to be a difficult task for many people, including professionals such as physicians. Realizing that (a) a conditional probability is not the same as its inverse and (b) base rates are relevant to the determination of posterior probabilities might help us avoid some of the difficulties. Translating the probabilities associated with base rates, sensitivity, and false positive rates into numbers of people, or "natural frequencies," also seems to help (e.g., Gigerenzer & Hoffrage, 1995). This is the approach used throughout this chapter. For example, in the first HIV problem, the sensitivity of 99.8% was translated into 1 HIV-infected individual who would test positive for HIV. Moreover, we translated the false positive rate of .01% into 1 HIV-free individual who would also test positive for HIV. This allowed us to easily determine that, out of the 2 individuals who tested positive for HIV, only 1—or 50%— would actually have the virus.

Gigerenzer and Hoffrage (1995) found that, when people are given information in a frequency format rather than a probability (or percentage) format, they find it easier to engage in Bayesian reasoning. They presented college students with Bayesian problems—including problems like the Blue cab problem and HIV problem considered above—in several different formats. When the participants were given frequencies, about half of their answers reflected proper Bayesian reasoning. When the students were given the same information in a probability format, less than one-third of their answers reflected proper Bayesian reasoning. Similarly, Hoffrage and Gigerenzer

(1998) asked German physicians to solve Bayesian problems—presented in either a frequency format or probability format— involving medical test results similar to the HIV problem presented at the beginning of the chapter. Only 10% of the physicians in the probability-format condition were able to give a correct answer, but 46% of those in the frequency-format condition answered correctly. Thus, using the approach illustrated in this chapter might help you improve your ability to engage in Bayesian reasoning.

Bayesian Reasoning in Legal Settings

As you might have inferred from the cab problem, Bayesian reasoning can also be useful in legal settings. Statistics that reflect conditional probabilities and base rates are becoming increasingly common in court settings (e.g., Thompson, 1989). For instance, imagine that you are a juror for a murder trial. The victim, a 29-year-old woman, was stabbed to death in her home. The defendant is a 30-year-old man who is a neighbor of the victim. After questioning a number of the victim's neighbors, police brought the defendant in for further questioning when they noticed he had a cut on his arm. During the trial, after you hear most of the prosecution's evidence against the defendant, you believe the evidence is very weak. You estimate there is only about a 10% chance that the defendant is guilty; this is the *prior probability* of guilt, which is used like a base rate in Bayesian calculations. However, at the last minute, the prosecution introduces evidence they recently obtained from a laboratory test of blood found at the crime scene. They state that DNA tests performed on the blood indicate that it has a genetic profile that matches the defendant's profile. The prosecution states that this genetic profile is found in only 1 person in 100, or 1% of the population. Thus, there is only a 1% chance that a randomly selected, innocent person's DNA would match the DNA found at the crime scene (i.e., false positive rate = 1%). Now that you know about this new evidence, what would you estimate is the probability that the defendant is guilty?

Let's approach this problem using the same intuitive approach to Bayesian reasoning used with previous problems. Consider 1,000 defendants in murder trials similar to the one just described, with similar evidence. Without the DNA evidence, you judged the prior probability of guilt to be 10%; thus, only 100 of these 1,000 defendants would actually be guilty. Let's assume that for 100% of these 100 guilty defendants, DNA tests would show that their DNA matches the DNA found at the crime scenes (i.e., sensitivity of DNA test = 100%). Thus, all 100 guilty defendants would be labeled "guilty" by the DNA test results. In other words, the number of true positives is 100. Of the 900 remaining defendants who are innocent, 1% of them—or 9 people—would have a matching, or "guilty," DNA profile. So, the number of false positives is 9. Out of a total of 109 people who were identified as "guilty" by the DNA test, 100 of them actually would be guilty. Therefore, the probability that a defendant is guilty given a matching DNA profile is 92% (i.e., 100/109 = 92%). As you can see, the DNA evidence should cause you to drastically increase your estimate of the probability of the defendant's guilt from 10% to 92%.[3]

A DNA test result is not the only type of evidence presented in criminal trials in which Bayesian reasoning might help a juror reach a decision. Laboratory tests conducted to determine if a defendant's blood type, hair, or carpet fibers match those found at the crime scene could be used by prosecutors in ways similar to DNA evidence (e.g., Thompson & Schumann, 1987). In each case, a juror would have to combine this evidence with his or her judgment of the prior probability of the defendant's guilt in order to arrive at the probability of guilt given the "matching" evidence.

Conditional Probability and the Monty Hall Dilemma

Let's move away from the realm of medical and legal settings for a moment and examine another phenomenon—one involving conditional probabilities—that often confuses people. Consider a hypothetical game run by a con artist in a major city. He shows you three small boxes on a table, one of which contains $10 (this fact is verified by another person among those gathered to watch this game). You are told that you will win the $10 if you choose the correct box, and you choose one of the boxes—Box A. Next, the con artist asks you, "What is the probability that you chose the box that contains $10?" You correctly answer "one-third." The con artist then opens Box C, which you did not choose, revealing there is nothing inside. He says, "Now you know the $10 is in either the box you chose or the unopened box; now what is the probability that the box you chose is the correct box?" You answer "50%." "Okay," says the con artist, "you can pick either of the two unopened boxes. Just put $10 on top of the box you choose. If it is the correct box, you can keep your $10 and the $10 in the box. But if you choose an incorrect box, I get to keep your $10." Believing you have a 50-50 chance of winning $10 by sticking with your original choice, you agree to put $10 on top of the box you initially chose. You then learn that you did not choose the correct box, and you lose your $10. "Oh well," you say to yourself, "I knew there was a 50% chance that I would lose." Was this just bad luck, or were the odds of winning not quite as favorable as you thought?

This hypothetical game is based on a famous, but defunct, television game show named *Let's Make a Deal*, hosted by Monty Hall. Imagine you are a contestant on this show. Monty asks you to choose among three doors: One of them has the keys to a

---

[3]Technically, the DNA evidence is not evidence of guilt; it is evidence that the defendant's blood was at the scene of the crime. It is possible that the defendant's blood was at the scene of the crime even though he is innocent. For example, if the defense proved that the defendant accidentally cut himself while working at the victim's house the day before the murder, this might negate the value of the DNA evidence. A similar defensive strategy was used by O. J. Simpson's lawyers to discount the value of some of the DNA evidence against him (see Thompson, 1996).

new car behind it, and behind the other two doors are goats. You choose Door A. Then, Monty opens one of the unchosen doors (Door C) and shows you there is a goat behind it. He invites you to stick with the door you chose or switch to Door B. Believing there is now a 50-50 chance that you chose the correct door, you decide to stick with Door A. Monty opens the door you chose and, to your dismay, a goat stands behind it. Was this just bad luck, or were the odds of winning not quite as favorable as you thought?

Actually, in both the Monty Hall situation (referred to as the **Monty Hall dilemma** by Granberg & Brown, 1995) and the con game, you would have been better off switching to the option you did not initially choose. Contrary to most people's intuition, the probability of winning by sticking with your initial choice is only 1/3, not 1/2. Thus, if you switch to the other option, the probability of winning becomes 2/3. Although some complex explanations of this fact have been offered, it is not difficult to understand—but you need to think in terms of conditional probabilities.

In the Monty Hall situation, you can either choose the correct door or an incorrect door. Because there are three doors and only one of them is correct, the probability of choosing the correct door is 1/3 and the probability of choosing an incorrect door is 2/3. After you make your selection, Monty reveals a goat behind one of the unchosen doors. He always reveals a goat, regardless of the door you choose, because he knows which door hides the keys to the new car. Thus, if you initially picked the correct door (probability = 1/3) you would win by sticking with your choice and lose if you switched to the door Monty did not open. However, if you initially chose an incorrect door (probability = 2/3) you would lose by sticking with your choice and win if you switched to the door Monty did not open. Therefore, if you played this game a number of times, the strategy of sticking with your initial choice would produce success only 1/3 of the time; the strategy of switching would produce success 2/3 of the time. In other words, the conditional probability of winning *given* a sticking strategy is 1/3, and the conditional probability of winning *given* a switching strategy is 2/3. Nevertheless, Granberg and Brown found that the vast majority of people to whom they presented this dilemma chose to stick with their original answer. As they explained, "People are especially prone to error when faced with problems involving conditional probability, and that is what is under consideration in the [Monty Hall dilemma]" (p. 712).

Because the game show no longer exists, perhaps the Monty Hall dilemma is more of an intellectual curiosity than something that has real-world applications. However, as implied with the con artist example, a con artist could make money from people's reasoning errors in this situation. If 3,000 people played this con game over the course of a few months, and all of them used a sticking strategy, an average of 2,000 (or 2/3) of them would lose and only 1,000 (or 1/3) would win. The losers would end up paying the con artist a total of $20,000, and the winners would cost him a total of $10,000; thus, his net gain would be $10,000. That is not a bad profit for a few months of "work."

Summary

Situations in our everyday lives may require understanding conditional probabilities and Bayesian reasoning. A conditional probability is the probability of some event, A, given that another event, B, has occurred. For example, you might want to know the probability that you have cancer given that a medical test indicated that you have it. A proper assessment of this probability requires that you integrate information about the base rate of the disease, the sensitivity of the test, and the false positive rate of the test. Bayes' theorem can be used to determine how to do this. We considered an intuitive approach to Bayesian reasoning that involved using these three pieces of information to determine the number of true positives and false positives that might be expected after testing a group of people. The probability that you have the disease given a positive test result can then be determined by simply dividing the number of true positives by the total number of true positives and false positives.

Research that has investigated people's ability to engage in Bayesian reasoning has found that they generally have a lot of difficulty with it; even physicians who have been trained to interpret the results of medical tests have difficulty with Bayesian reasoning. One reason why people have difficulty with Bayesian reasoning is confusion over the meaning of conditional probabilities; sometimes they assume that a conditional probability is the same as its inverse. This confusion could result in a person assuming that the probability of a disease given a positive test result is the same as the probability of a positive test result given the disease. Another reason that people have difficulty with Bayesian reasoning is base-rate neglect. For example, a physician might assume that the base rate of a disease is irrelevant to the probability that a person has the disease given a positive test result. But, as we have seen, the base rate can have a dramatic impact on that probability. One way to improve your ability to engage in Bayesian reasoning is to translate the probabilities associated with base rates, sensitivity, and false positive rates into numbers of people, or "natural frequencies." People, including physicians, who are presented with information in this "frequency format" show an improved ability to engage in Bayesian reasoning.

Finally, Bayesian reasoning is also useful in legal settings. For example, in a criminal trial, Bayesian reasoning could be used to integrate one's initial assessment of the defendant's guilt with evidence of a match between the defendant's hair and hair samples found at the scene of the crime.

● PRACTICE PROBLEMS

1. During her yearly medical checkup, a woman decides to have a mammogram to make sure she is free from breast cancer. The result of the test indicates that she has cancer in one of her breasts. While discussing the result with her doctor, she learns that the rate of breast cancer among women like her is only about 1 in 1,000. She also learns that (a) when a woman has breast cancer, a mammogram detects it 90% of the time and (b) when a woman does not have breast cancer, a positive mammogram occurs 10% of the time. What is the probability that this woman actually has breast cancer given the positive mammogram?

2. A young man tests positive for HIV, and he tells his physician, "It seems highly unlikely that I have HIV; I don't have any of the risk factors." He asks the physician, "What percentage of people like me have HIV?" The physician responds, "I'm not sure, but it doesn't really matter how prevalent HIV is in the population of people like you. Population statistics are irrelevant to individual cases. If the test showed that you have HIV, then you have it." What specific error in Bayesian reasoning is the physician making?

3. A woman serving as a juror in a rape trial initially believes, based on the prosecution's evidence, that there is about a 20% chance that the defendant is guilty of this crime. She then hears that the prosecution found some of the perpetrator's hair at the scene of the crime, and tests revealed that it matches the defendant's hair. In addition, she learns that this type of hair is found in about 30% of the male population. What is the probability that the defendant is guilty given this new evidence?

4. One night while watching the local news on television, you see a reporter interview a local store owner who decided to ban teenagers from her store. She explained to the reporter that she heard that teenagers are responsible for 75% of the thefts that occur in the area's stores. That prompted her to ban teenagers from her store because, as she explained, "I figured that if I let teenagers in my store, there is a very good chance that they would steal something from me." What specific error in reasoning is this store owner making?

Conclusion

If you took a statistics course in college, you probably learned how to calculate and interpret statistics such as probabilities, estimated population values, correlations, regression values, and factorial ANOVAs. At the time, you may have asked yourself why you needed to learn statistics, especially if your desired career did not involve performing statistical analyses. One answer commonly given to this question is that some professions (e.g., counseling or social work) require reading the research literature—that is, articles that contain statistical analyses—to stay informed about the latest treatments or theoretical developments. Another common answer is that we encounter statistics in our daily lives, and we need to be able to understand this information. For example, we often see statistics in media reports about crimes, deaths, the weather, and political polls.

This book provides an additional answer to the question of why it is useful to learn statistics: because many of the concepts one learns in a statistics course are useful for making decisions and judgments in everyday life. This book has shown how statistical concepts can help us improve our decision-making and critical-thinking skills. Perhaps you never imagined that statistical concepts such as probability, the law of large numbers, estimation, correlation, regression, and Bayes' theorem had any relevance to your everyday life. If so, then I hope this book has given you a greater appreciation for the value of training in statistics.

This book also examined a number of difficulties people have with statistical reasoning and the role of heuristics and biases in producing some of these difficulties. Perhaps you have already begun to witness examples of these difficulties in your daily life. For instance, you or someone you know may have (a) attached too much significance to what was probably a chance event, (b) made an ill-advised generalization based on a small sample or a single case, (c) witnessed regression toward the mean and generated some other explanation for it, or (d) allowed confounding variables to cloud your judgment. If you have not witnessed these things, keep your eyes peeled; you will probably see them soon. I hope that making you more aware of these difficulties will help you become less susceptible to them in your own life.

Finally, if you are a student of psychology, perhaps you have also begun to see some connections between statistics and other areas of psychology that you did not see before reading this book. As implied throughout the book, the statistical concepts we discussed can enhance our understanding of a variety of phenomena (e.g., gambler's fallacy, correspondence bias, false consensus, and stereotypes) that students often learn about in cognitive and social psychology courses. Of course, this book is by no means a complete discussion of how statistical concepts are related to other areas of psychology or of how statistical concepts are related to everyday life. Now that you have learned about some of these connections, perhaps you will begin to see connections that go far beyond our discussion.

Glossary

Anchor-adjustment heuristic a mental shortcut in which we estimate a value by first anchoring on an initial estimate and then adjusting the estimate—often insufficiently—from that point.

Analysis of covariance (ANCOVA) a statistical procedure used to evaluate differences among two or more groups while statistically controlling for the effects of unwanted variables.

Analysis of variance (ANOVA) a statistical procedure used to evaluate differences among two or more groups.

Attribution the act of determining the causes behind behaviors or events.

Availability heuristic a mental shortcut in which we estimate the probability or frequency of an event based on how easily we can imagine or recall it happening.

Base-rate neglect the tendency underuse base-rate, or population, information and favor case-specific information.

Bayesian reasoning reasoning that reflects an understanding of Bayes' theorem.

Bayes' theorem a rule that expresses the relationships among various conditional probabilities.

Bystander effect the tendency for people to be less likely to help a person in need when other people (i.e., bystanders) are nearby at the time of the emergency.

Conditional probability the probability of some event, A, given that another event, B, has occurred.

Confounding variable an uncontrolled variable that acts along with the independent variable to influence an outcome.

Conjunction a combination of two or more events.

Conjunction fallacy the tendency to assign a higher probability to a combination of two events than to one of those events.

Conjunction rule a rule of probability that states that the probability of a combination of two events cannot be greater than the probability of one of those events.

Correlation an association, or relationship, between two events or variables.

Correlation coefficient a statistic that indicates the degree of the relationship between two variables.

Correspondence bias the tendency to draw inferences about a person's dispositions from behavior that can be completely explained by situational factors.

Estimation the use of sample statistics to estimate population parameters.

False consensus effect the tendency to believe that others share our opinions and habits.

False positive rate the percentage of individuals incorrectly identified as having some characteristic or disease.

Fundamental attribution error *see* **Correspondence bias**.

Heuristics mental shortcuts or rules of thumb.

Hot hand belief the belief that a basketball player (for example) is more likely to make a shot after making one or more shots in a row than after missing one or more shots in a row.

Gambler's fallacy the belief that a long run of a particular outcome of a chance event (e.g., a coin landing on tails) makes a different outcome more likely on the next try.

Illusory correlation the perception of a relationship that does not exist or a relationship that is not as strong or in the same direction as one believes.

Illusion of control belief that we have more control over an event than we actually do.

Illusion of transparency tendency to overestimate the degree to which our internal states are noticed by others.

Interaction occurs when the effect of one independent variable depends on another independent variable.

Law of large numbers (LLN) the fact that larger samples provide better estimates of populations than do smaller samples.

Main effect the effect of one independent variable.

Monty Hall dilemma a situation in which a person tries to guess which of three doors hides a prize, after which a host reveals that one of the unchosen doors does not contain the prize. Then the person is given the option of switching to the remaining unchosen door.

Multiplication rule a rule used to find the probability of the conjunction of two or more independent events by multiplying together their individual probabilities.

Negative correlation a relationship between variables in which they change in opposite directions; as one increases, the other decreases.

Overjustification effect a reduction in a person's intrinsic (or internal) interest in an activity after being given extrinsic (or external) rewards for engaging in that activity.

Person-who statistics relying on a single case to contradict the implications of group statistics.

Placebo effect a change in a person's condition that results from the person's belief or expectation that such a change would take place.

Positive correlation a relationship between variables in which they change in the same direction; as one increases, the other also increases.

Positive-test strategy searching for evidence that matches the hypothesis being tested, as opposed to evidence that would match an alternative hypothesis.

Posterior probability the probability that you actually have some characteristic or disease (for example), given that a test indicated you have it.

Prior probability the base rate, or percentage, of individuals in a population who possess some characteristics or disease.

Random sample a sample in which each member of the population had an equal chance of being included in the sample.

Regression toward the mean the tendency for extreme values on one variable to be accompanied by less extreme values on a second variable when the two variables are imperfectly correlated.

Representativeness heuristic a mental shortcut in which we judge probability based on the degree to which one thing resembles, or is representative of, another thing.

Sampling distribution a distribution of sample values from all possible random samples of a certain size.

Sensitivity the percentage of individuals correctly identified by a test as possessing some characteristic or disease (among those who actually possess it).

Spotlight effect the tendency to overestimate the extent to which other people notice our behaviors and appearance.

Stereotype a belief—which is usually inaccurate or overgeneralized—about the characteristics shared by members of some group.

Subjective probability a subjective estimate of the probability of some event.

Wells effect the psychological difference between saying, "There is an X% chance that this is true" and saying, "This is true based on evidence that is X% reliable."

References

Note: Page numbers in parentheses at the end of each reference indicate where the citation can be found in this book.

Alloy, L. B., & Tabachnik, N. (1984). Assessment of covariation by humans and animals: The joint influence of prior expectations and current situational information. *Psychological Review, 91,* 112–149. (pp. 59, 63, 68)

Anderson, C. A., & Anderson, D. C. (1984). Ambient temperature and violent crime: Test of the linear and curvilinear hypotheses. *Journal of Personality and Social Psychology, 46,* 91–97. (p. 61)

Around the NFL. (January 4, 2000). *The Cincinnati Enquirer,* p. D2. (p. 85)

Bar-Hillel, M. (1973). On the subjective probability of compound events. *Organizational Behavior and Human Performance, 9,* 396–406. (p. 14)

Berkowitz, L. (1983). Aversively stimulated aggression: Some parallels and differences in research with animals and humans. *American Psychologist, 38,* 1135–1144. (p. 96)

Borgida, E., & Nisbett, R. E. (1977). The differential impact of abstract vs. concrete information on decisions. *Journal of Applied Social Psychology, 7,* 258–271. (p. 33)

Brekke, N., & Borgida, E. (1988). Expert psychological testimony in rape trials: A social-cognitive analysis. *Journal of Personality and Social Psychology, 55,* 372–386. (p. 119)

British crime rise blamed on alcohol. (July 19, 2000). *The Cincinnati Enquirer,* p. A16. (p. 95)

Bushman, B. J., & Cooper, H. M. (1990). Effects of alcohol on human aggression: An integrative research review. *Psychological Bulletin, 107,* 341–354. (p. 96)

Campbell, D. T., & Kenny, D. A. (1999). *A primer on regression artifacts.* New York: The Guilford Press. (pp. 75, 87)

Casscells, W., Schoenberger, A., & Graboys, T. B. (1978). Interpretation by physicians of clinical laboratory results. *The New England Journal of Medicine, 299,* 999–1001. (pp. 2, 114, 116)

Chapman, L. J., & Chapman, J. P. (1967). Genesis of popular but erroneous psychodiagnostic observations. *Journal of Abnormal Psychology, 72,* 193–204. (pp. 66–68)

Chapman, L. J., & Chapman, J. P. (1969). Illusory correlation as an obstacle to the use of valid psychodiagnostic signs. *Journal of Abnormal Psychology, 74,* 271–280. (pp. 66–68)

Chun, W., & Lee, H. (1999). Effects of the difference in the amount of group preferential information on illusory correlation. *Personality and Social Psychology Bulletin, 25,* 1463–1475. (p. 70)

Combs, B., & Slovic, P. (1979). Causes of death: Biased newspaper coverage and biased judgments. *Journalism Quarterly, 56,* 837–843, 849. (pp. 11, 45)

Darley, J. M., & Latané, B. (1968). Bystander intervention in emergencies: Diffusion of responsibility. *Journal of Personality and Social Psychology, 8,* 377–383. (p. 98)

Dawes, R. M. (1988). *Rational choice in an uncertain world.* San Diego: Harcourt Brace Jovanovich. (p. 117)

Dawes, R. M., Faust, D., & Meehl, P. E. (1989). Clinical versus actuarial judgment. *Science, 243,* 1668–1674. (p. 82)

DeGowin, E., & DeGowin, R. (1969). *Bedside diagnostic examination* (2nd ed.). New York: Macmillan. (p. 119)

Derry, S., Levin, J. R., & Schauble, L. (1995). Stimulating statistical thinking through situated simulations. *Teaching of Psychology, 22,* 51–57. (p. 1)

Dollard, J., Doob, L., Miller, N., Mowrer, O. H., & Sears, R. R. (1939). *Frustration and aggression.* New Haven, CT: Yale University Press. (p. 96)

Eagly, A. H., & Steffen, V. J. (1984). Gender stereotypes stem from the distribution of women and men into social roles. *Journal of Personality and Social Psychology, 46,* 735–754. (p. 108)

Eddy, D. M. (1982). Probabilistic reasoning in clinical medicine: Problems and opportunities. In D. Kahneman, P. Slovic, & A. Tversky (Eds.), *Judgment under uncertainty: Heuristics and biases* (pp. 249–267). Cambridge, England: Cambridge University Press. (pp. 118–119)

Federal Bureau of Investigation. (1998). *Crime in the United States.* Washington, DC: U.S. Government Printing Office. Retrieved December 13, 1999, from the World Wide Web: http://www.fbi.gov/ucr.htm. (p. 9)

Fiedler, K. (1991). The tricky nature of skewed frequency tables: An information loss account of distinctiveness-based illusory correlations. *Journal of Personality and Social Psychology, 60,* 24–36. (p. 70)

Fiedler, K., Walther, E., & Nickel, S. (1999). The auto-verification of social hypotheses: Stereotyping and the power of sample size. *Journal of Personality and Social Psychology, 77,* 5–18. (pp. 64, 69)

Fong, G. T., Krantz, D. H., & Nisbett, R. E. (1986). The effects of statistical training on thinking about everyday problems. *Cognitive Psychology, 18,* 253–292. (pp. 1, 3, 30)

Fridson, M. S. (1993). *Investment illusions.* New York: Wiley. (p. 19)

Galton, F. (1886). Regression towards mediocrity in hereditary stature. *Journal of the Anthropological Institute of Great Britain and Ireland, 15,* 246–263. (p. 88)

Gardner, M. (1972). Mathematical games: Why the long arm of coincidence is usually not as long as it seems. *Scientific American, 227,* 110–112B. (p. 20)

Gigerenzer, G., & Hoffrage, U. (1995). How to improve Bayesian reasoning without instruction: Frequency formats. *Psychological Review, 102,* 684–704. (pp. 2, 114–115, 120)

Gigerenzer, G., Hoffrage, U., & Ebert, H. A. (1998). AIDS counseling for low-risk clients. *AIDS Care, 10,* 197–211. (pp. 2, 113–114, 116, 119)

Gilbert, D. T., & Malone, P. S. (1995). The correspondence bias. *Psychological Bulletin, 117,* 21–38. (pp. 48, 100)

Gilovich, T. (1991). *How we know what isn't so: The fallibility of human reason in everyday life.* New York: The Free Press. (p. 37)

Gilovich, T., Medvec, V. H., & Savitsky, K. (2000). The spotlight effect in social judgment: An egocentric bias in estimates of the salience of one's own actions and appearance. *Journal of Personality and Social Psychology, 78,* 211–222. (p. 56)

Gilovich, T., Savitsky, K., & Medvec, V. H. (1998). The illusion of transparency: Biased assessments of others' ability to read one's emotional states. *Journal of Personality and Social Psychology, 75,* 332–346. (p. 54)

Gilovich, T., Vallone, R., & Tversky, A. (1985). The hot hand in basketball: On the misperception of random sequences. *Cognitive Psychology, 17,* 295–314. (p. 37)

Granberg, D., & Brown, T. A. (1995). The Monty Hall dilemma. *Personality and Social Psychology Bulletin, 21,* 711–723. (p. 123)

Gutierres, S. E., Kenrick, D. T., & Partch, J. J. (1999). Beauty, dominance, and the mating game: Contrast effects in self-assessment reflect gender differences in mate selection. *Personality and Social Psychology Bulletin, 25,* 1126–1134. (p. 46)

Hamill, R., Wilson, T. D., & Nisbett, R. E. (1980). Insensitivity to sample bias: Generalizing from atypical cases. *Journal of Personality and Social Psychology, 39,* 578–589. (pp. 50, 120.)

Hamilton, D. L., & Gifford, R. K. (1976). Illusory correlation in interpersonal perception: A cognitive basis of stereotypic judgments. *Journal of Experimental Social Psychology, 12,* 392–407. (p. 69)

Hamilton, D. L., & Sherman, J. W. (1994). Stereotypes. In R. S. Wyer, Jr., & T. K. Srull (Eds.), *Handbook of social cognition* (2nd ed.). Hillsdale, NJ: Erlbaum. (p. 70)

Heath, L., & Petraitis, J. (1987). Television viewing and fear of crime: Where is the mean world? *Basic and Applied Social Psychology, 8,* 97–123. (p. 57)

Hoffrage, U., & Gigerenzer, G. (1998). Using natural frequencies to improve diagnostic inferences. *Academic Medicine, 73,* 538–540. (pp. 2, 114, 120–121)

Hoover, D. W., & Milich, R. (1994). Effects of sugar ingestion expectancies on mother-child interactions. *Journal of Abnormal Child Psychology, 22,* 501–515. (p. 68)

Horn, D. (April 7, 2000). *The Cincinnati Enquirer,* pp. A1, A14. (p. 13)

Hoyert, D. L., Kochanek, K. D., & Murphy, S. L. (1999). Deaths: Final data for 1997. *National Vital Statistics Reports, 47*(19), 99–1120. Retrieved December 15, 1999, from the World Wide Web: http://www.cdc.gov/nchs/releases/99facts/99sheets/97mortal.htm. (p. 10)

Huesmann, L. R. (1998). The role of social information processing and cognitive schema in the acquisition and maintenance of habitual aggressive behavior. In R. G. Green & E. Donnerstein (Eds.), *Human aggression: Theories, research, and implications for policy.* New York: Academic Press. (p. 96)

Jackson, L. A., Hunter, J. E., & Hodge, C. N. (1995). Physical attractiveness and intellectual competence: A meta-analytic review. *Social Psychology Quarterly, 58,* 108–122. (p. 69)

Jennings, D. L., Amabile, T. M., & Ross, L. (1982). Informal covariation assessment: Data-based versus theory-based judgments. In D. Kahneman, P. Slovic, & A. Tversky (Eds.), *Judgment under uncertainty: Heuristics and biases* (pp. 211–230). Cambridge, England: Cambridge University Press. (p. 71)

Kahneman, D., & Tversky, A. (1972). Subjective probability: A judgment of representativeness. *Cognitive Psychology, 3,* 430–454. (p. 28)

Kahneman, D., & Tversky, A. (1973). On the psychology of prediction. *Psychological Review, 80,* 237–251. (pp. 83–84)

Kelley, H. H. (1967). Attribution theory in social psychology. In D. Levine (Ed.), *Nebraska symposium on motivation.* Lincoln: University of Nebraska Press. (p. 97)

Kelley, H. H. (1973). The processes of causal attribution. *American Psychologist, 28,* 107–128. (p. 97)

Kenrick, D. T., Gutierres, S. E., & Goldberg, L. L. (1989). Influence of popular erotica on judgments of strangers and mates. *Journal of Experimental Social Psychology, 25,* 159–167. (p. 46)

Kenrick, D. T., Neuberg, S. L., Zierk, K. L., & Krones, J. M. (1994). Evolution and social cognition: Contrast effects as a function of sex, dominance, and physical attractiveness. *Personality and Social Psychology Bulletin, 20,* 210–217. (p. 46)

Klayman, J., & Ha, Y-W. (1987). Confirmation, disconfirmation, and information in hypothesis testing. *Psychological Review, 94,* 211–228. (p. 64)

Krueger, J., & Clement, R. W. (1997). Estimates of social consensus by majorities and minorities: The case for social projection. *Personality and Social Psychology Review, 1,* 299–313. (p. 43)

Kruger, J., Savitsky, K., & Gilovich, T. (1999). Superstition and the regression effect. *The Skeptical Inquirer, 23,* 24–29. (pp. 86–87)

Kulig, J. W. (2000). Effects of forced exposure to a hypothetical population on false consensus. *Personality and Social Psychology Bulletin, 26,* 629–636. (p. 53)

Kunda, Z. (1999). *Social cognition: Making sense of people.* Cambridge, MA: The MIT Press. (pp. 63–64)

Kunda, Z., & Nisbett, R. E. (1986). The psychometrics of everyday life. *Cognitive Psychology, 18,* 195–224. (p. 72)

Langer, E. (1975). The illusion of control. *Journal of Personality and Social Psychology, 32,* 311–328. (p. 70)

Latané, B., & Darley, J. (1970). *The unresponsive bystander: Why doesn't he help?* New York: Appleton-Century-Crofts. (p. 100)

Latané, B., & Rodin, J. (1969). A lady in distress: Inhibiting effects of friends and strangers on bystander intervention. *Journal of Experimental Social Psychology, 5,* 189–202. (p. 55)

Lepper, M. R., Greene, D., & Nisbett, R. E. (1973). Undermining children's intrinsic interest with extrinsic reward: A test of the overjustification hypothesis. *Journal of Personality and Social Psychology, 28,* 129–137. (p. 96)

Lichtenstein, S., Slovic, P., Fischhoff, B., Layman, M., & Combs, B. (1978). Judged frequency of lethal events. *Journal of Experimental Psychology: Human Learning and Memory, 4,* 551–578. (p. 10)

Lord, C. G., Lepper, M. R., & Preston, E. (1984). Considering the opposite: A corrective strategy for social judgment. *Journal of Personality and Social Psychology, 47,* 1231–1243. (p. 66)

Marsh, H. W. (1984). Students' evaluations of university teaching: Dimensionality, reliability, validity, potential biases, and utility. *Journal of Educational Psychology, 76,* 707–754. (p. 34)

Meehl, P. E. (1954). *Clinical versus statistical prediction: A theoretical analysis and review of the evidence.* Minneapolis: University of Minnesota Press. (p. 82)

Milgram, S. (1963). Behavioral study of obedience. *Journal of Abnormal and Social Psychology, 67,* 371–378. (p. 99)

Milgram, S. (1974). *Obedience to authority.* New York: Harper and Row. (p. 99)

Miller, A. G., Gillen, B., Schenker, C., & Radlove, S. (1974). The prediction and perception of obedience to authority. *Journal of Personality, 42,* 23–42. (p. 100)

Miller, D. T., Turnbull, W., & McFarland, C. (1989). When a coincidence is suspicious: The role of mental simulation. *Journal of Personality and Social Psychology, 57*, 581–589. (p. 23)

Myers, D. G. (2001). *Psychology* (6th ed.). New York: Worth. (pp. 20, 96, 117)

National Safety Council. (1991). *Accident facts.* Chicago: Author. (p. 12)

Niedermeier, K. E., Kerr, N. L., & Messe, L. A. (1999). Jurors' use of naked statistical evidence: Exploring bases and implications of the Wells effect. *Journal of Personality and Social Psychology, 76*, 533–542. (p. 23)

Nisbett, R., & Borgida, E. (1975). Attribution and the psychology of prediction. *Journal of Personality and Social Psychology, 32*, 932–943. (pp. 99, 120)

Nisbett, R., Krantz, D. H., Jepson, C., & Kunda, Z. (1983). The use of statistical heuristics in everyday inductive reasoning. *Psychological Review, 90*, 339–363. (pp. 3, 88.)

Nisbett, R., & Ross, L. (1980). *Human inference: Strategies and shortcomings of social judgment.* Englewood Cliffs, NJ: Prentice-Hall. (pp. 25, 33, 63–64, 71, 82, 85, 93, 95–96)

Page, E., & Keith, T. (1981). Effects of U.S. private schools: A technical analysis of two recent claims. *Educational Researcher, 10*, 7–17. (p. 108)

Paulos, J. A. (1988). *Innumeracy: Mathematical illiteracy and its consequences.* New York: Hill and Wang. (pp. 13, 14, 20, 26, 28)

Paulos, J. A. (1994, March). Counting on dyscalculia. *Discover, 15*, 30–36. (p. 37)

Pilcher, J. (2000, July 5). Buses fuller, air cleaner. *The Cincinnati Enquirer,* p. A1, A9. (p. 86)

Redelmeier, D. A., & Tversky, A. (1996). On the belief that arthritis pain is related to the weather. *Proceedings of the National Academy of Sciences of the United States of America, 93*, 2895–2896. (p. 73)

Richins, M. L. (1991). Social comparison and the idealized images of advertising. *Journal of Consumer Research, 18*, 71–83. (pp. 45–46)

Ross, L. (1977). The intuitive psychologist and his shortcomings. In L. Berkowitz (Ed.), *Advances in experimental social psychology* (Vol. 10). San Diego, CA: Academic Press. (p. 48)

Ross, L., Amabile, T. M., & Steinmetz, J. L. (1977). Social roles, social control, and biases in social-perception processes. *Journal of Personality and Social Psychology, 35*, 485–494. (p. 47)

Ross, L., Greene, D., & House, P. (1977). The "False Consensus Effect": An egocentric bias in social perception and attribution processes. *Journal of Experimental Social Psychology, 13*, 279–301. (p. 52)

Ross, M., & Sicoly, F. (1979). Egocentric biases in availability and attribution. *Journal of Personality and Social Psychology, 37*, 322–336. (p. 49)

Schaller, M. (1992a). In-group favoritism and statistical reasoning in social inference: Implications for formation and maintenance of group stereotypes. *Journal of Personality and Social Psychology, 63*, 61–74. (p. 108)

Schaller, M. (1992b). Sample size, aggregation, and statistical reasoning in social inference. *Journal of Experimental Social Psychology, 28*, 65–85. (p. 105)

Schaller, M., Asp, C. H., Rosell, M. C., & Heim, S. J. (1996). Training in statistical reasoning inhibits the formation of erroneous group stereotypes. *Personality and Social Psychology Bulletin, 22*, 829–844. (pp. 105, 109)

Schaller, M., & O'Brien, M. (1992). "Intuitive analysis of covariance" and group stereotype formation. *Personality and Social Psychology Bulletin, 18*, 776–785. (pp. 103, 107, 109)

Sherman, S. J., Presson, C. C., Chassin, L., Corty, E., & Olshavsky, R. (1983). The false consensus effect in estimates of smoking prevalence. *Personality and Social Psychology Bulletin, 9,* 197–207. (p. 53)

Silka, L., & Albright, L. (1983). Intuitive judgments of rate change: The case of teenage pregnancies. *Basic and Applied Social Psychology, 4,* 337–352. (p. 110)

Slovic, P., Fischhoff, B., & Lichtenstein, S. (1982). Facts versus fears: Understanding perceived risk. In D. Kahneman, P. Slovic, & A. Tversky (Eds.), *Judgment under uncertainty: Heuristics and biases* (pp. 463–489). Cambridge, England: Cambridge University Press. (p. 12)

Smedslund, J. (1963). The concept of correlation in adults. *Scandinavian Journal of Psychology, 4,* 165–173. (p. 63)

Snyder, M., & Cantor, N. (1979). Testing hypotheses about other people: The use of historical knowledge. *Journal of Experimental Social Psychology, 15,* 330–342. (p. 64)

Snyder, M., & Swann, W. B. (1978). Hypothesis-testing processes in social interaction. *Journal of Personality and Social Psychology, 36,* 1202–1212. (p. 65)

Stanovich, K. E. (2001). *How to think straight about psychology* (6th ed.). Needham Heights, MA: Allyn & Bacon. (pp. 19, 33, 95, 108)

Stice, E., & Shaw, H. E. (1994). Adverse effects of the media portrayed thin-ideal on women and linkages to bulimic symptomatology. *Journal of Social and Clinical Psychology, 13,* 288–308. (p. 46)

Thompson, S. C. (1999). Illusions of control: How we overestimate our personal influence. *Current Directions in Psychological Science, 8,* 187–190. (p. 70)

Thompson, W. C. (1989). Are juries competent to evaluate statistical evidence? *Law and Contemporary Problems, 52,* 9–41. (p. 121)

Thompson, W. C. (1996). DNA evidence in the O. J. Simpson trial. *University of Colorado Law Review, 67,* 827–857. (p. 122)

Thompson, W. C., & Schumann, E. L. (1987). Interpretation of statistical evidence in criminal trials. *Law and Human Behavior, 11,* 167–187. (p. 122)

Tversky, A. & Kahneman, D. (1971). Belief in the law of small numbers. *Psychological Bulletin, 76,* 105–110. (p. 36)

Tversky, A., & Kahneman, D. (1973). Availability: A heuristic for judging frequency and probability. *Cognitive Psychology, 5,* 207–232. (p. 11)

Tversky, A., & Kahneman, D. (1974). Judgment under uncertainty: Heuristics and biases. *Science, 185,* 1124–1131. (p. 15)

Tversky, A., & Kahneman, D. (1982a). Evidential impact of base rates. In D. Kahneman, P. Slovic, & A. Tversky (Eds.), *Judgment under uncertainty: Heuristics and biases* (pp. 153–160). Cambridge, England: Cambridge University Press. (pp. 116, 118–119)

Tversky, A., & Kahneman, D. (1982b). Judgments of and by representativeness. In D. Kahneman, P. Slovic, & A. Tversky (Eds.), *Judgment under uncertainty: Heuristics and biases* (pp. 84–98). Cambridge, England: Cambridge University Press. (p. 24)

Tversky, A., & Kahneman, D. (1983). Extensional versus intuitive reasoning: The conjunction fallacy in probability judgment. *Psychological Review, 90,* 293–315. (pp. 15, 16, 17)

Vorauer, J. D., & Claude, S. (1998). Perceived versus actual transparency of goals in negotiation. *Personality and Social Psychology Bulletin, 24,* 371–385. (p. 55)

Ward, W. C., & Jenkins, H. M. (1965). The display of information and the judgment of contingency. *Canadian Journal of Psychology, 19,* 231–241. (p. 63)

Wells, G. L. (1992). Naked statistical evidence of liability: Is subjective probability enough? *Journal of Personality and Social Psychology, 62,* 739–752. (p. 21)

Wyatt, W. J., Posey, A., Welker, W., & Seamonds, C. (1984). *The Skeptical Inquirer, 9,* 62–66. (p. 20)

Yip, P. S. F. (1998). Age, sex, marital status and suicide: An empirical study of east and west. *Psychological Reports, 82,* 311–322. (p. 117)

Index

Analysis of covariance (ANCOVA)
 confounding variables and, 104
 explanation of, 104
 intuitive ANCOVA and everyday life, 105–111
 perceptions of changes in subpopulations and, 109–111
 stereotyping and, 106–109
Analysis of variance (ANOVA)
 attribution theory and, 97–100
 everyday reasoning and, 95–100
 factorial ANOVA, 94
 interaction, 94
 main effect, 94
 single vs. multiple causes, 95–97
 two-factor and single-factor designs, 94
Attribution
 definition of, 47
 fundamental attribution error/correspondence bias, 48, 98
 theory, 97–98

Base rate, 113
 neglect, 118–120
Bayes' Theorem, 115
 Bayesian reasoning in legal settings, 121–122
 common difficulties in Bayesian reasoning, 116–120
 improving Bayesian reasoning, 120–121
Bystander effect, 55

Conditional probability
 Bayes' theorem and, 114–116
 confusing, 117–118
 definition, 114
 Monty Hall dilemma and, 122–123

Confounding variable, 104
Conjunctions, 13–18
 conjunction fallacy, 16–18
 conjunction rule, 17
Correlation
 assessing in everyday life, 61–63
 biased expectations and illusory correlations, 66–68
 coefficient, 61
 definition of, 61
 difficulties in assessing, 63–70
 illusion of control and, 70–71
 positive-test strategy, 64–66
 positive vs. negative, 61
 stereotypes and, 69–70
Correspondence bias, 48, 98

Estimation
 attribution and, 47–49
 biased samples from the mass media and, 45–47
 generalizing from the self to others, 51–57
 process of, 44–45
 single cases and, 49–51

False consensus effect, 52–53
False positive rate, 113
Fundamental attribution error, 48, 98

Gambler's fallacy, 35

Heuristics, 7
 anchor-adjustment, 14, 54, 56
 availability, 11
 representativeness, 17, 28–29
Hot hand, 37

Illusion of control, 70–71
Illusion of transparency, 53–55
Illusory correlations, 66–70
Interaction, 94

Law of large numbers
   explanation of, 26–28
   intuitive version of, 29–31
   law of small numbers and, 36
   perceptions of randomness and, 35–37
   person-who statistics, 33–35
   randomness and belief in the hot hand, 37–39
   representativeness heuristic and, 28–29
   single cases vs. group statistics, 32–33

Monty Hall dilemma, 122–123
Multiplication rule, 8

Overjustification effect, 96

Person-who statistics, 33–35
Placebo effect, 32
Positive-test strategy, 64–66
Posterior probability, 114

Probability
   assessing risk, 8–13
   chance events, 18–21
   coincidences, 19–21
   conjunction rule, 17
   multiplication rule, 8
   prior, 113
   research and, 7–8

Random sample, 44
Regression
   correlation and, 76–79
   equation, 77
   everyday examples of, 84–88
   law of large numbers and, 88–89
   linear regression, 76
   regression toward the mean, 79–81

Sampling distributions, 27
Sensitivity, 113
Spotlight effect, 55–57
Stereotype, 69
Subjective probabilities, 8

Wells effect, 23

TO THE OWNER OF THIS BOOK:

We hope that you have found *Everyday Statistical Reasoning: Possibilities and Pitfalls* useful. So that this book can be improved in a future edition, would you take the time to complete this sheet and return it? Thank you.

School and address:_____

Department:_____

Instructor's name:_____

1. What I like most about this book is:_____

_____

_____

2. What I like least about this book is:_____

_____

_____

3. My general reaction to this book is:_____

_____

4. The name of the course in which I used this book is:_____

_____

5. Were all of the chapters of the book assigned for you to read?_____

    If not, which ones weren't?_____

6. In the space below, or on a separate sheet of paper, please write specific suggestions for improving this book and anything else you'd care to share about your experience in using the book.

_____

_____

_____

_____

Optional:

Your name: _____ Date: _____

May Wadsworth quote you, either in promotion for *Everyday Statistical Reasoning* or in future publishing ventures?

Yes: _____ No: _____

Sincerely,

Timothy J. Lawson

FOLD HERE

BUSINESS REPLY MAIL
FIRST CLASS   PERMIT NO. 358   PACIFIC GROVE, CA

POSTAGE WILL BE PAID BY ADDRESSEE

ATT: *Timothy J. Lawson*

Wadsworth–Thomson Learning
511 Forest Lodge Road
Pacific Grove, California 93950-9968

NO POSTAGE
NECESSARY
IF MAILED
IN THE
UNITED STATES

FOLD HERE